A Concise Survey of
Animal Behavior

"Watch Geese"
Anthony Benson

A Concise Survey of Animal Behavior

ERIKA K. HONORÉ
Department of Food Animal and Equine Medicine
North Carolina State University
College of Veterinary Medicine
Raleigh, North Carolina

PETER H. KLOPFER
Department of Zoology
Duke University
Durham, North Carolina

ACADEMIC PRESS, INC.
Harcourt Brace Jovanovich, Publishers
San Diego   New York   Boston
London   Sydney   Tokyo   Toronto

ACADEMIC PRESS, INC.
San Diego, California 92101

United Kingdom Edition published by
Academic Press Limited
24–28 Oval Road, London NW1 7DX

ISBN   0-12-355065-3

PRINTED IN THE UNITED STATES OF AMERICA
90 91 92 93   9 8 7 6 5 4 3 2 1

Contents

Illustrations

and intensity (loudness) by the darkness of the lines.

Foreword: Research in Animal Behavior—So You Want to Learn More?

A prerequisite for studying animals is a fundamental love and curiosity about them. Students of animal behavior come from all walks of life, but the one thing they share is a fascination for the habits of animals, a fascination usually developed at an early age from watching pets, wildlife in the backyard, or animals at the zoo.

Two of the most important traits of a successful student of animal behavior are a keen eye for observation and the ability to understand animal behavior from the animal's point of view. The keen eye can look at an animal—a gazelle at the zoo for example—and see more than just an animal; the keen observer notices the posture of the gazelle, how the gazelle is holding its head, and what facial expression it has. The gazelle's behavior could be interpreted in two ways: the anthropomorphic interpretation might be that the gazelle, if its head is held high, is looking for a friend; the more accurate interpretation would be that the gazelle is alert. (In the wild gazelles must stay alert for predators.) The best interpretation of an animal's behavior unfolds when one considers the animal's natural environment. In the natural environment, an animal's fitness depends upon its ability to avoid predators, and to find food and mates. When interpreted from this perspective, rather than an anthropomorphic one, greater clarity is achieved.

How is research in animal behavior conducted?

Any general rules or theories of animal behavior arise from many observations. Two independent observations of the same occurrence do not constitute development of a new theory on this particular occurrence. Two observations of a gorilla beating its chest when it saw a blond-haired girl looking into its zoo exhibit does not mean that the presence of blond-haired girls will always stimulate chest-beating in gorillas. To be able to make such a generalization one would have to observe this phenomenon several more times *and* determine that there were no other possible causes. Likewise, if you visited the zoo and noticed that each time you walked by the elephant exhibit the elephants were standing in one corner, you could not justly conclude that they prefer that corner of their enclosure over all others. You could only draw such a conclusion if you had made hundreds of observations of the elephants at different times of the day and in varying weather conditions.

In order to test predictions about animal behavior, researchers do the same thing that any other scientist would do: employ the scientific method. The first step is to ask a question and then formulate a prediction or hypothesis. Usually questions arise from preliminary observations. For instance, in a large black-headed gull colony, Niko Tinbergen observed that the parent birds at a nest carefully removed all pieces of egg shell from the nest after the chicks had hatched. His question was, "Why do gulls remove the egg shells from the nest?" His prediction was that the bright white lining of the egg shells, in contrast to the mottled brown outside of the egg shell, was very conspicuous to predators of gull eggs and chicks. Unlike the outside of the egg, which is mottled brown in color and well camouflaged in the nest by its coloration, the bright white lining of the shell is highly visible against the brown nest's background. When the egg shells were removed, the nest became less visible to over-flying predators.

The second step in the scientific method is to design a method to test your prediction. The important characteristics of the method are that it must be a reproducible method and that it must be repeated several times. The method is only valid if it can be reproduced by someone else under similar conditions. No shortcuts or new methods can be incorporated at will, there must be one carefully designed method. The method of experiment must be repeated several times so that researchers can use statistical tests to determine whether the results are really meaningful and not due to random chance. To test his prediction about egg shells attracting predators to the nest, Tinbergen placed 120 eggs in an area adjacent to the gull colony. These eggs were all slightly concealed by straw; 60 had egg shells next to them and 60 did not.

The next step involves actually collecting the data. Tinbergen recorded how many of the 120 eggs were taken by predators.

The final step of the scientific method involves analyzing and interpreting the data. In Tinbergen's experiment, only 13 of the eggs that did not have the shell next to them were taken, compared to 39 of the 60 eggs that did have shells lying within 5 cm of them. Tinbergen concluded that the near presence of an egg shell helped predators to find concealed prey. Therefore, the brood would be endangered if the egg shells were not carried away.

There are also behavioral questions that can be addressed without doing experimental manipulations like Tinbergen did. In the zoo setting, we often ask questions about how changes in an enclosure will affect an animal's behavior. The method for addressing this question involves recording the animal's behavior before and after the change, and then analyzing the results in a "before" versus "after" comparison.

Who Conducts Behavioral Research? Can Anyone Do It?

Most scientists in the field of animal behavior have doctoral degrees in psychology, biology, or zoology. This rigorous training provides a strong foundation in theory and experimental design. Designing research studies requires correct selection and usage of the various observation methods. Such design should be left to the experts, but the collection of data often requires hundreds of hours. If a study is well designed and the objectives and methodology are clear and unambiguous, then it is possible to train nonbiologists and nonpsychologists to collect the data with a high level of accuracy. It is not necessary to have a doctorate to participate in this phase of animal behavior research.

Often research scientists involve both undergraduate and graduate students in the data collection phase of their projects. At Zoo Atlanta we have gone one step further and enlisted, in addition to students, the assistance of volunteers interested in animal behavior. A requirement for such volunteer positions is patience, dedication, and a keen eye for observation. Anyone who studies animals knows that animals do not always behave as you expect or wish them to. Resting and sleeping are part of an animal's behavioral repertoire and sometimes fill an entire sampling period; but, with a little patience and persistence, several components of a specie's behavioral repertoire may be observed.

What Kind of Behavioral Research Goes on at a Zoo?

Zoos all over the world have housed animals for centuries, but only recently have they begun to realize the value of, and potential for, research within their own facilities. Zoo Atlanta is a good example of a zoo that has gone through the transition from the old philosophy of holding as many animals as possible behind bars in sterile concrete cages, to the new philosophy of exhibition combined with conservation—exhibition of animals in naturalistic habitats and in natural social groupings.

One important area of animal behavior research in the zoo relates to housing animals in naturalistic habitats and the evaluation of a specie's behavior within its exhibit. The purpose of this research is to promote species-typical behavior within the exhibit space. Why is this important? Zoo visitors gain a much better understanding and appreciation for wildlife when they see the animals behaving as they would in nature. In addition, animals enjoy a much higher level of psy-

chological well-being when they are housed in an enclosure
that offers the complexity and environmental and social stimuli they would encounter in a wild setting.

Knowing that gorillas live in social groups consisting of
an adult male and a group of females and their offspring,
many zoos have begun housing their gorillas in such family
groups. Providing a natural social group is the first step toward offering the appropriate environment to animals in captivity. The second step is offering an enclosure, or habitat,
which has several of the physical components of the animal's
natural environment. Zoo Atlanta has provided its four family groups of gorillas with large outdoor enclosures and has
evaluated the behavior of the animals within these enclosures. The gorilla groups were initially slow to explore their
new outdoor environments, preferring to stay close to familiar objects and surroundings (Ogden, 1989). Ogden found that
the components of the new outdoor environment that elicited
natural behavior and were preferred by the gorillas were flat
areas with rocks and trees and logs. Ogden concluded that the
most important feature of any habitat is that it should offer
the animals choices and options for different types of activity.
The next step in this process of promoting normal active behavior is to continue to add to the exhibit, evaluating each addition along the way.

Similarly, exhibiting and maintaining orangutans is
often a challenge—their overwhelming response to the captive environment is lethargy and inactivity. Stimulating natural behavior in this species is, in addition to the reasons mentioned above, essential for successful reproduction. In a study
of several orangutan groups housed in nine zoos, Perkins
(1989) determined that, in general, there are four features of
an enclosure that are most likely to promote normal active
behavior in orangutans: (1) the volume of the enclosure, (2)
the usable surface area (i.e., climbing structures), (3) the number of movable objects, and (4) the number of animals in the
exhibit. These findings thus provide the foundation for guid-
ing us in enhancing the captive, naturalistic environments of
orangutans.

Natural, active behavior in animals is largely composed

of behavior associated with food-finding. One major difference between the wild and the captive settings is the amount of time devoted to foraging. In the wild, most animals devote a large portion of time to looking for food, preparing the food (i.e., peeling bark and leaves off of trees), and finally eating it. In captivity, food is commonly served already prepared, chopped into bite-size pieces, and presented practically on a silver platter. Animals thus spend no time looking for, or preparing, the food and probably a very minimal amount of time actually eating the food. One obvious method of promoting more species-typical behavior, and at the same time enriching the animal's experience in its enclosure, is to try to increase foraging time. Feeding branches of fresh leaves to gorillas and other herbivores like gazelles, giraffes, and zebras offers a nutritious treat, while also increasing the amount of time the animal spends processing the food, pulling each individual leaf off of the branch.

Feeding enrichment studies have also been conducted on bears. The most vivid memory of bears at the zoo is, more often than not, one of single animals pacing back and forth across their exhibits. This pacing is referred to as a stereotypy, which means that the pacing behavior is very rigid—the bear paces exactly the same number of steps each time without variation. Stereotypic behavior is most often seen as a response to boredom, and so it is a goal of zoos to enhance the psychological well-being of the bears by trying to discourage pacing. Bears normally occur solitarily in the wild and they spend a substantial proportion of their time foraging for food. Once again, serving food in bite-size pieces all at once leaves the bears with a lot of extra time in the day—time that would otherwise be devoted to foraging in the wild. Increasing their foraging time in captivity has successfully been achieved through hiding food all over the exhibit to encourage the bears to move around and search for their food and by offering "fishcicles," Zoo Atlanta's method of increasing the amount of time it takes a bear to eat a fish. This has even been shown to lead to a decrease in pacing.

In addition to these types of animal behavior research focusing on the behavior of animals in their captive environments, zoo research has contributed greatly to the body of knowledge of little-known and endangered species. Often, very little is known about the behavior and reproductive cycles of animals until they can be closely studied in captivity. The drill, for instance, is an endangered species of baboon from West Africa. It is closely related to the largest of the African baboons, the mandrill. The distribution and status of the drill in the wild are poorly understood at best. Although the mandrill breeds readily in captivity, the closely related drill does not. Through comparative studies of the two species at Zoo Atlanta, we hope to determine the reproductive cycle and factors that will influence reproductive behavior. Success in reproduction in captivity could eventually lead to the reintroduction of drills to the wild.

It is the responsibility of zoos to discourage collecting animals from the wild and, instead, to try and maintain their own self-propagating populations. An example of an animal that is known to have a poor record of breeding success in captivity, and is taken in great numbers from the wild, is the flamingo. None of the six species of flamingos is endangered, however the sporadic breeding success of several colonies of flamingos could lead to a threatened status. The factors that distinguish good years of flamingo chick production from bad years are not clearly understood. Without understanding exactly what triggers breeding in flamingos, we will be left helpless when wild flamingos simply cease reproducing.

It seems that successful reproduction in flamingos requires more than just putting a male and a female together. Flamingos live in large flocks for a reason—the large number of individuals perform a series of courtship displays, or "group" displays, during the breeding season that facilitate pair-bond formations and stimulate breeding. Research in the zoo setting has shown that high levels of group display activity correlate with increased breeding behavior. Now the search is on to determine which environmental and social factors influence, or trigger, the group displays.

Each of the studies mentioned has been accomplished through hours of observation by trained observers, many of whom did not have formal education in animal behavior. It is possible for animal behavior enthusiasts to indulge in animal behavior just by visiting their local zoos. A little more probing might turn up a research project and a willingness to train volunteers in data collection as we do at Zoo Atlanta. There is still a lot you can learn without traveling to exotic and far away places.

Beth Franke Stevens
Research Biologist
Zoo Atlanta
Atlanta, Georgia

Preface

A. To Whom Are We Speaking?

Animal behavior, otherwise known as ethology, has become an important element of courses in anthropology, psychology, veterinary medicine, and animal science, but is not yet deemed sufficiently vital to warrant a full term's course. There exist many texts on ethology, ranging from good to indifferent to bad, but all are alike in being designed for full semester courses. Even our favorite text (Klopfer, 1974), which is among the shortest, does not fit into the time usually reserved for animal behavior. We therefore discuss the insights and generalizations most important for an understanding of animal behavior and how much is encompassed by the field. Examples about particular groups of animals facilitate easy cross-referencing. Even though we principally address ourselves to our professional colleagues, we believe this book will be interesting and useful to the sophisticated pet owner and hobbyist as well as to the serious student.

B. What About Jargon?

It is a curious trait of *Homo sapiens* that we consider puzzling phenomena explained, once they have been named. In any given discipline, therefore, specialized terms serve as proof of common understanding. Ethology is no different. Consider, for example, the term *displacement*. If two conflicting tendencies are simultaneously stimulated, such as flight and attack in the presence of an intruder, a third, seemingly irrelevant, behavior pattern may appear: the animal may sleep, or pull at a stem of grass, or eat. This irrelevant activity is called a displacement and is thereby explained. The term, of course, is loaded: it carries with it considerable theoretical baggage. In this instance, the theory in question is Lorenz's hydraulic model. The supposition is that the fluids or energy specific to fighting and the fluids or energy specific to fleeing block one another's release valves, requiring the accumulating fluids (energies) to slop over into some other reservoir or pathway. This is an intriguing suggestion, easy to visualize and understand. Unfortunately, the notion of action-specific energies for fighting is conjectural, and there is much to say in opposition to the notion; in the end, displacement may really be only another fable. We might say that a gull threatened by an intruder with whom it does not seem to want to fight, or from whom it appears unable to flee, indulges in displacement preening. We have, however, merely described a situation and explained very little. Indeed, would our description suffer if the word displacement were stricken from the sentence?

Jargon is a fact of life, however, and while we may vow personally to eschew it, we cannot avoid learning it. Thus the more important instances are listed in the index.

The first edition of *An Introduction to Animal Behavior: Ethology's First Century* (Klopfer and Hailman, 1967), provided us with the inspiration for this volume. Thus, even though J. P. Hailman was far distant when this book was written, we remain indebted to him for his ideas and stimulation.

We thank Dale Peele for her efforts on our sometimes balky word processor and her patience with our always difficult handwriting, and Rachel Simon, for her skillful editing. Thanks also to Gerard Honoré, Peter Kappeler, and Beth Franke Stevens for their photographic contributions.

Erika Honoré
Peter Klopfer

An Introduction to Animal Behavior: Ethology's First Century. Second Edition, P. H. Klopfer. Prentice Hall, 1974.

A Concise Survey of
Animal Behavior

1

Historical Hinterland

A. Folk Knowledge and Rome's Geese

When the geese of Rome honked an alarm at a barbarian's intrusion, they were doing what was expected of them. The Romans knew their geese, and recognized that the excited honking indicated the presence of trespassers. This sort of folk knowledge has characterized humans since our ancestors set out to hunt the first rabbit or tame the first cat. Indeed, no hunter could hope for sucess without fairly intimate and accurate ideas of the behavior of his prey.

Of course, this does not preclude some fantasy. Until this century, many bird watchers, watching the silhouettes of nocturnal birds against the moon, were convinced that the moon was the migrants' destination during autumnal flights. Where else could they possibly go during the cold winter months? The list of fanciful notions could be endlessly extended, but to do so might lead to embarrassment: not all the fantasies are merely historical oddities, whose silliness we can appreciate.

We have made some progress in understanding animal behavior, and, quite apart from hunters, our forebears were not entirely unsophisticated. In medieval paintings, the animals appear in appropriate poses and in the proper habitat. Geese are not shown in forests, or swans in pastures.

Even today, people who do not know the term ethology still practice it. So-called primitives, or as we now call them, preliterate cultures, demonstrate a fine-tuned knowledge of the animals important to them. Some African herders, for instance, insert hollow rods into the vaginas of their cows and then, before milking, blow through the rods. The tribesmen know that this facilitates extracting milk, although they probably do not realize that the cervical stimulation causes a reflex release of oxytocin, the milk let-down hormone. Such knowledge must be regarded as sophisticated, even when the mechanism (the oxytocin reflex) is not known.

B. From Aristotle to Darwin and Company

The Greek philosophers, foremost among them Aristotle, articulated and formalized what was then the state of the art in animal behavior. Aristotle did not always have his facts right; nonetheless, much of what he said has a modern ring. For about 2000 years, from Aristotle to Darwin, almost nothing important was added to our understanding of animals, and a great deal of foolishness abounded. In eighteenth century England, for instance, a pig that escaped into a neighbor's garden could be tried and hanged for a deliberate act of trespass (Klopfer and Polemics, 1989).

With the flowering of science in the next century, however, natural history and particularly behavior studies came into their own. *Ethology*, a term that previously applied to the portrayal of human characters on a stage, was redefined as "The study of animals, not as corpses reeking with formaldehyde ... but as living things in their natural habitat (Jaynes, 1969, p. 602).

Even with the changed definition and studies then being done, the new term did not immediately catch on. When its use did become widespread, in the mid-nineteenth century, it was largely identified with the writings of ardent Lamarckians, who believed in the inheritance of acquired traits. The biologist Alfred Giard, as well as others, popularized both the term and Lamarck's views on how individual adaptations to the environment influenced the appearance, and hence evolution, of offspring. As Lamarck's views lost ground, so did the use of ethology. Not until the 1940s did the term regain any measure of popularity, and then it was almost exclusively associated with animal behavior studies carried out in the manner of Konrad Lorenz. Only since the 1960s has ethology come to have a more general definition as the study of animal behavior, especially using naturalistic methods, as opposed to methods entailing surgical intervention or confinement to highly artificial environments. In other words, the goal of ethology is to understand the natural behavior of animals.

As the term ethology went through many changes over the centuries, so too did the field itself. Before Darwin, the study of behavior had the practical goals of improved husbandry or improved sport. Indeed, breeders of racing pigeons contributed in no small way to the development of Darwin's views on the power of selective breeding in shaping a species. However, with the acceptance by many biologists of Darwin's evolutionary models, the focus of ethologists shifted. They became concerned with the issues of the evolution of mental capacities and the laws describing this evolution.

Foremost among the new, evolution-oriented breed of ethologists was the Englishman Lloyd Morgan. He disdained as sterile the controversies over whether a given behavior pattern was innate or acquired, conscious or unconscious. He developed experimental methods that precluded consideration of factors not under the experimenter's control. Morgan's canon advises experimenters to never use a complex explanation if a simpler one will do.

A similar attitude pervaded the work of Jacques Loeb, who pioneered physiological (or mechanistic) approaches to the study of behavior. Neither man had patience for vague, nondemonstrable concepts such as consciousness.

A host of other workers was also active in the last half of the nineteenth century. Among the most important are the prodigious Herbert Spencer, who contributed a volume on the evolution of behavior, along with reams on every other conceivable topic; Charles Darwin, who wrote several books that dealt explicitly with the evolution of behavior; George John Romanes, Darwin's successor in ethology; and the sagacious William James, generally regarded as the founder of comparative psychology, who, it was said, was as good a novelist as his brother Henry was a psychologist. William James did, in fact, do much more than leave us pleasantly written tracts; indeed, he is the most modern of the nineteenth century ethologists. Fortunately, several good (and short) reviews of the history of this period are available.

C. Twentieth Century Beginnings

As the nineteenth century ended, behavior studies in the United States were largely characterized by their experimental nature. Homogeneous populations were selected as subjects (most often inbred mice and pigeons), environments were simplified, and conditions were systematically varied. But the Skinner box (an automated device by which rats or pigeons were trained to peck or press bars in response to particular signals) did not appear in 1900, and even then a great many Americans were continuing to study behavior in the naturalistic tradition. Some were truly amateurs, such as the author–explorer Ernest Thomas Seton, whose descriptions of wildlife behavior beguile all who read him, or the ornithologist Margaret M. Nice. Others, such as Charles Whitman and William Wheeler, set standards of clarity, originality, and precision that continue to serve us today, even though their work is more readily classified as naturalistic than experimental. J. B. Watson, the behavioristic psychologist, began his scientific career with naturalistic studies. His work on bird behavior with K. S. Lashley, later the founder of modern neurobiology, was as important a contribution as were his later theories on how behavior can be controlled by the environment.

But even with some naturalistic practitioners, the United States became, for the most part, the home of the vigorous, laboratory-centered study of behavior. Usually this was under the aegis of comparative psychologists whose goal was less to understand the behavior of their subjects than to trace the course of mental evolution.

By contrast, in Europe and England the naturalistic tradition continued to dominate. The mechanistic approach of the laboratory scientists originated in the physiology laboratories of Germany. The dominant figures in twentieth century Europe, however, were indisputably the naturalists, from Julian Huxley through David Lack in England, Oscar Heinroth to Konrad Lorenz and Niko Tinbergen on the continent (though they performed experiments, too). Some scientists, such as Wolfgang Köhler with his studies of reasoning in apes, or Karl von Frisch with the dancing bees, combined the two traditions so completely as to defy classification.

Yet the Old World and the New did differ, although the major difference was less in method than in goal. For the psychologists of the United States, it was to understand the structure and rules of behavior; for their Old World counterparts, it was to examine issues of adaptive purpose and ecological significance. For instance, U.S. research would investigate how the frequency of rewards alters behavior, whereas the typical Old World research would ask why stomachaches are more effective punishment than food deprivation.

The Old and New World traditions are more similar today; indeed, a distinction is scarcely possible, given the numerous exchanges between countries of students, professors, and publications. But why was this not the case in 1920? Why did few U.S. authors cite their European counterparts, or even know of their work, and vice versa? William Wheeler was described by his fellow American ethologists as the modern scientist best suited to converse with Aristotle, so broad was the range of his intellect. His European contemporaries were not less well educated or intelligent. Perhaps the personal attributes of leading spokesmen contributed to their separation. For instance, Watson, among the most prolific of the American workers, was as charismatic (and dogmatic) as his European opposite a decade later, Konrad Lorenz. Science, it seems, is no less influenced by personality and passion than by ideas and reason.

D. Suggested Readings

A concise and chronologic history of ethology, with references to original sources and recent, more detailed, accounts
is to be found in P. H. Klopfer's *An Introduction to Animal
Behavior: Ethology's First Century* (Englewood Cliffs, New
Jersey: Prentice-Hall, 1974). A personal account of the princi-
pals in the development of modern ethology is W. H. Thorpe's
Origin and Rise of Ethology (New York, New York: Praeger,
1979).

2

Current Concepts

A. Introduction

If a discipline comes of age when it achieves departmental rank at even the stodgiest academic institutions, then ethology's birth must be dated in the 1960s. Christening day presumably came a decade later when the Nobel Prize was awarded to Niko Tinbergen, Konrad Lorenz, and Karl von Frisch for their seminal contributions to the new discipline.

Niko Tinbergen stated that the mission of ethologists was to answer these four questions about behavior:

1. What is its function (adaptive purpose)?
2. What is its history (evolution)?
3. How is it initiated and controlled (mechanism)?
4. How does it change or develop (ontogeny)?

Consider a species of the family of dancing flies, the *Empeidae*. The males and females engage in an extensive courtship, during which the male spins, and then presents to the female, a parcel of silk. How did this habit ever get started? What tells the male how to do it? Does he improve his performance with practice? These sorts of queries, argues Tinbergen, represent the domain of and challenge for ethologists.

There is no simple way to classify or order the multitudes of studies that address these questions. Many workers deal with several of them, utilizing one particular animal. Tinbergen did most of his research, using the above questions, with a single species of gull. Other researchers concentrate on one question and may utilize different animals. Both approaches offer advantages and drawbacks. Concentrating on one question can lead to excessive dependence on one experimental design, approach, or apparatus, with the investigator losing sight of the forest for the trees. Focusing on one animal, on the other hand, can result in only superficial answers to some of the four questions.

We will, in the next few sections, attempt to organize the current concepts in ethology around Tinbergen's four questions. This is an arbitrary classification; indeed, since scarcely any one person is able to address all four questions with a variety of animals, individual perspectives are bound to be limited. Unless ethologists are also committed and voracious readers, their outlook is likely to be about as balanced as that of the proverbial blind men palpating the elephant. Thus, the reader should be aware that this presentation mirrors the approaches of most ethologists.

B. Adaptation

Voltaire's anti-hero, Pangloss, suffered innumerable misfortunes, as did his companions. These included, at one time, the loss of part of a buttock. These stalwarts never complained, for they had been convinced that this was the best of all possible worlds; hence, apparent disabilities must also be advantageous—adaptations, no less.

Our views are perhaps less sanguine than those of Pangloss, but we, as biologists, do share his inclination to assume that all features of an organism, its structure and actions, serve some useful purpose. In the eyes of traditional biologists, committed to the Darwinian paradigm, purpose need not entail conscious design. What it does mean is that the organ or behavior in question has been retained in the species because it contributes to greater fitness. Those individuals possessing the trait are going to be able to produce a proportionately greater share of the individuals of subsequent generations than those lacking the trait.

For instance, if large antlers make a stag more successful at attracting and inseminating females than do small antlers, then, other things being equal, large-antlered males will account for more of the next generation and may ultimately displace small-antlered rivals altogether. It should be remembered, however, that the greater energetic costs of building and maintaining large antlers might ultimately outweigh the benefits. Thus, it can be an error to assume that any conspicuous character must be advantageous. The question, "What does this trait do and what purpose does it serve?" must not presuppose that the trait is an adaptation.

It has been shown that large antlers, up to a certain point, do provide more benefits than costs, and ethologists have enumerated and calculated these. But, for behavior (or any physical trait) to be considered an adaptation in the Darwinian or evolutionary sense, it must meet several criteria. It must be heritable (natural selection or fitness differences cannot exist for nonheritable features); there must exist alternative traits (if antlers can only be of one size, there is no point in speaking of large antlers as adaptive); and, finally, it must be possible to demonstrate how and in what manner the trait contributes to greater fitness than its alternative.

Certain gulls, as Niko Tinbergen (1958) once described, remove the shell fragments from around their nests once their young have been hatched. Is this behavior an adaptation? What would happen if the shell fragments were not removed? Tinbergen was able to show that the pieces of broken shell served to attract crows, which will avidly devour young gull chicks. The chicks of gulls that removed the shells were more likely to escape predation by crows during parental fishing forays.

Adaptations may have more effects than meet the eye. Eggshell removal in the gulls is a behavioral response that

helps protect the young. It need not have further consequences. Increased antler growth, on the other hand, as an
adaptation for sexual rivalry, also has a direct impact on the
stags' nutritional requirements and metabolism.

Adaptations for a social life are likely to be even more
complex and convoluted in their effects. Animals living in
colonies, such as ground squirrels, will often give a particular
call signaling the presence of danger—in the case of the squirrels, aerial predators. When the alarm sounds, all squirrels
scurry to their burrows. Was the alarm call adaptive? Surely
not for the animal who gave it; had that animal remained
silent, it could have reached its own burrow faster and reduced its probability of capture. Possibly, by not warning others, this squirrel could have assured the hawk or other predator a meal at the expense of its neighbor. Thus, when it
re-emerged from its burrow it might, until the last meal had
been ingested, have been immune from further attacks. So
why warn others? In fact, specialized warning calls are such a
conspicuous feature of social species that it is hard not to assume they are adaptive. This must mean they benefit the animal performing the act. But how?

Several explanations are possible, but we will only explore one. It is based on a now famous model developed by W.
H. Hamilton (1963). He re-emphasized an old truism that the
more closely related two individuals are, the more genetic
material they hold in common. Two full siblings share about
50% of their genes; cousins, 25%; and so on. Two of your
brothers represent as much of you as you do yourself, and
three brothers are half again as much as you. Thus, your fitness or your personal contribution to gene pools of the future,
is increased by 50% if by sacrificing yourself you can assure
the survival of three brothers, as compared to your brothers
dying and you surviving.

This is a gross simplification, of course, for trade-offs are
rarely so extreme, and must be measured as probabilities.
However, it can be mathematically shown (and empirically
demonstrated) that if your alarm call provides a benefit to
enough of your relatives, it may be adaptive even if it increases your chance of being a hawk's dinner. This important
principle has become known as *kin-selection* and provides

the basis for much of biosociology. For now, the point is that the function of alarm calls, and much similar behavior, can only be understood in the context of the social milieu in which it occurs.

C. Evolution

It is not altogether facetious to ask whether evolution is inevitable. When the question first occurred to us, it was as a result of reading the play *R.U.R.* by the Czech playwright Capek. *R.U.R.*, which stands for Rossum's Universal Robots, was a comedy written when the word "robot" was still new. The play concerned automatons shaped to resemble human beings and designed to perform certain human functions.

In the play, Rossum is an engineer who designs a robot that is able to assume some of the more arduous and unpleasant jobs normally performed by men and women. As time passes, Rossum perfects this design so that robots come increasingly to replace human workers until there is scarcely anything but procreation left to *Homo sapiens*. But the human procreative capability is shown not to be a unique feature; the robots are able to staff their own assembly lines and reproduce their own kind, assembling them, albeit with occasional mistakes, from the parts salvaged from old, worn-out robots.

Homo sapiens' undoing comes when the robots develop sentience. To a degree, this was desired and planned for by the engineer. A robot with feelings can avoid hazardous situations; if you do not feel pain, you are likely to destroy yourself in any number of situations from which you might otherwise retreat. Having programmed these robots with some minimal degree of feeling, however, the engineer found that, unavoidably, the robots also came to harbor other more subtle emotions (geneticists recognize this as a pleiotropic effect), including resentment of being exploited by their human masters. As these feelings intensified, the robots rose in rebellion and conspired to destroy their masters and the enterprise they had built. It was a case of "Robots of the world—unite! You have nothing to lose but your machinery."

In the final scene of the play, as the robots storm the citadel where the engineer and his colleagues are beseiged, the engineers desperately analyze their blueprints, searching for the aspect of their design that had allowed the robots to develop such ardent feelings. At the brink of despair, the engineers finally order one of their own robot servants to take another of its kind into the dissecting room so they can trace the actual circuitry once again. The robot to whom the command is given is built in the shape of a female, and upon hearing the command, it begs, "No, no! Don't order me to do this. Take me instead!" Simultaneously, the intended victim, a male-type robot, shouts, "No, no, don't listen to her! Take me! Take me!" The engineers stare at both robots and wonder whether this could really be Adam and Eve all over again. Of course, on this dramatic note, with the door beginning to crack from the weight of the robots outside, the play ends.

Was the robot's evolution inevitable? Consider the four characteristics with which Capek endowed his system, all sufficient and necessary. The first was the capacity for self-replication. The second was that this replication did not result in identical copies every time. The replication process was not error-free. That is, sometimes the robots made mistakes and their offspring differed in some particular way from the parents.

This second characteristic is very significant. Usually, a change in the structure of a well-designed instrument results in decreased effectiveness. This is, however, not always the case. If you have a well-constructed watch, any kind of random intervention—even a not-so-random intervention by an incompetent watchmaker—will likely produce a timepiece less accurate after the interference than before. But every once in a while, a crude shake or a drop or some other brutal treatment may improve the timepiece, especially if it did not function well before. To put it another way, as you focus a microscope, the more precisely you adjust the fine-focus screw, the more likely you are to get it into focus. But every once in a while, a random twist of the dial will do it. Indeed, the farther you are from a perfect focus to begin with, the more likely it is that a random twist will improve the sharpness of the image, rather than diminish it.

So, while errors may be expected to represent a net loss in precision of function or design, a finite probability does exist that there will be an improvement in some aspect of the instrument's function or design. Therein lies the importance of these errors in replication.

The third characteristic of the evolving robots was mortality. They wore out.

Fourth, and finally, each robot was constructed of components drawn from a finite store of parts. There cannot be an infinite number of machines.

These last two attributes are quite reasonable; if there were no limit to the number of robots around at any one time, then there would never be an advantage in the production of certain kinds of errors. The situation might be thought of in the following terms. Suppose we have a finite supply of nuts and bolts. The individual machines would compete for that store of parts, all drawing from it in order to build up replicas of themselves. After a certain number of replicas had been built, the original machines would break down, and their parts would be deposited in the central store. A finite store of parts would thus circulate through the pool of machines.

Suppose further that, as a result of a copy error, one of these robots were able to build replicas of itself a little more rapidly than others, or to build replicas using a smaller number of parts. That particular robot and those of its descendants who share that particular copy error would gradually monopolize the total store of parts. Competition among machines results from the fact that the resource (parts) on which the competitors depend is limiting.

Of course, this competition would not exist if the store of parts were infinite or if the robots were absolutely identical in their replication processes. But if copy errors occurred, then the robots would not be precisely identical. Eventually, one machine would be better in some way at replication than the others. If the store of parts were finite, then that mutant would gradually increase in frequency. That is to say, it and its progeny would monopolize the limiting resource.

Bear in mind what is involved here. A copy error produced a new model. The new model produced progeny, usually like itself (although, of course, through the generations new copy errors would occur and there would be yet a third model). Initially, after just the first copy error had occurred, if it were indeed an advantageous one, the progeny of that particular lineage would capture ever-increasing proportions of the parts from the storehouse. This would not be of any significance, if the storehouse were supplied with an infinite number of parts. So both attributes, limited supply and copy errors, are important.

Can any of this explain behavioral differences among animals? Ground-feeding birds often use their feet to feed, or at least to expose their prey. For example, in the eastern United States, fox sparrows hop with both feet, kicking leaves backwards; wood thrushes methodically stir leaves with one foot at a time; and thrashers move leaves with their curved bills, using their legs only to propel themselves. All three of these behavior patterns expose ground-dwelling insects, on which the birds feed. Are differences solely a matter of anatomic structures and peculiarities? An ethologist would certainly wish to find out.

As visitors to Oxford, England, we were bewildered by the seemingly random distribution of house numbers in the oldest parts of the town. Quite by chance, while perusing sixteenth- and seventeenth-century town maps, we came across a clue to the numbering system. Houses were numbered chronologically in their order of construction, not, as the post office might prefer it, according to their order along a street.

The explanation for Oxford addresses is an accident of history. It is possible that a similar answer serves for the three species of birds described above. How a bird feeds is perhaps of less consequence than that it does so efficiently. Possibly it was more efficient for the birds to concentrate on one particular movement and pattern than on a mix of sometimes kicking and sometimes hopping (how would they decide when to do which?). Additionally, perhaps the ancestral forms differed anatomically, so it was easier for the original fox sparrow to kick and the original thrush to stir. Unfortunately, we can only infer historical explanations; even the footprints left by the dinosaurs or the amber-embedded remains of insects provide but indirect evidence on the origins of a trait.

The realization that historical factors play a role in the appearance of present behavior does, however, have some significant implications. In the early twentieth century, a number of biologists came to realize that behavioral traits can, under some circumstances, be as *conservative* as the structures that serve them. (Conservative means that the traits do not change readily through time, either because the selective pressures that shaped them remain constant or because no countervailing pressures have arisen.)

The movements or displays performed during courtship among ducks are a good example. The courtship patterns are stereotyped and vary little among members of a species, but vary considerably between species. This serves to maintain interspecific reproductive fidelity, which was undoubtedly as advantageous in times past as now. O. Heinroth, a great German ornithologist and systematist, proposed that the courtship patterns of ducks and geese could thus prove to be as reliable indices of relatedness as structures of bone and feathers (Klopfer, 1974). In other words, the patterns can indicate the history of ducks to us.

Or consider again the dancing flies, described by the American T. C. Schneirla (1953). In one species the male secretes a mass of silk as he flies about the female, fashioning a bundle, which he presents to her. In another species, the male encases a fragment of leaf in a web and proffers it to the female. The male of another species wraps a small fly in a web, and that is his gift to the female, while in yet another species, the fly is simply presented unwrapped. These are just a few of the many variations on this behavioral trait.

Schneirla infers that in the ancestral species, mating males reduced the likelihood that they themselves would become prey to female appetites by presenting a killed fly as diversion. The appearance of the wrapped-prey ploy further enhanced the safety of the mating male, in that the female was distracted for a longer time by having to probe the silken envelope. The male that substituted plant fragments obviously found that it paid to proffer the empty mass of silk, a symbol of the ancestral prey. Imaginative as Schneirla's scheme may be, it nonetheless suggests how, step-by-step, simple functional acts may be evolutionarily transformed into those that seem bizarre and otherwise unexplainable.

A danger in evolutionary speculations lies in their circularity. If we already know, on the basis of other data, what a particular evolutionary sequence or relationship is, then we might be able to infer the historical sequence of changes in the behavior. If, however, we begin with the supposed sequence of changes and from them infer the evolutionary history of the species, which is what Heinroth sought to do for ducks, a great many assumptions have to be made.

Konrad Lorenz (1974) attempted to make these assumptions explicit in his Nobel address.

As a pupil of the comparative anatomist and embryologist Ferdinand Hochstetter, I had the benefit of a very thorough instruction in the methodological (sic) procedure of distinguishing similarities caused by common descent from those due to parallel adaptation. In fact, the making of this distinction forms a great part of the comparative evolutionist's daily work. Perhaps I should mention here that this procedure has led me to the discovery which I personally consider to be my own most important contribution to science. Knowing animal behavior as I did, and being instructed in the methods of phylogenetic comparison as I was, I could not fail to discover that the very same methods of comparison, the same concepts of analogy and homology, are as applicable to characters of behavior as they are in those of morphology. This discovery is implicitly contained in the works of Charles Otis Whitman and of Oscar Heinroth; it is only its explicit formulation and the realization of its far-reaching inferences to which I can lay claim. A great part of my life's work has consisted in tracing the phylogeny of behavior by disentangling the effects of homology and of parallel evolution. Full recognition of the fact that behavior patterns can be hereditary and species specific to the point of being homologizable was impeded by resistance from certain schools of thought, and my extensive paper on homologous motor patterns in Anatidae was necessary to make my point. (Lorenz, 1974, p. 231)

But, whence comes the knowledge that there has been common descent? In the absence of independent evidence, the argument is circular. Picture the forelimbs of bats, whales, sea lions, moles, and humans. Are these resemblances analogous, (similar in function), or are these similarities homologous, (similar in ancestry)? Lorenz says the latter is the case, because "the very dissimilarity of their functions makes it extremely improbable that the manifold resemblances of their forms could be due to ... analogy" (Lorenz, 1974, p. 231). But on what basis does Lorenz conclude that functions are dissimilar? How do we define dissimilarity anyway? Is the difference mostly one of scale and perspective?

The problems of distinguishing analogy and homology are old, complex, and far less simple than Lorenz asserts. Nonetheless, the beguiling simplicity of evolutionary explanations continues to attract us, encouraging us to use them to explain intricate or unwelcome traits. M. Mead (personal communication, 1964) once remarked that the advantage of evolutionary explanations lay in almost always being able to find an example to explain any trait. A telling example is the developmental consequences of rearing infants with foster parents in place of natural parents. Some psychologists once argued against Federal support for day-care centers on the grounds that foster mothering had unfortunate consequences. This conclusion was based on studies of maternally deprived rhesus monkeys, *Macaca mulatta*. Had the witnesses cited studies instead with the closely related pig-tailed monkeys, *M. nemestrina*, they would have come to quite different conclusions.

Certainly, Lorenz is correct that there are differences in the ease or speed with which traits can be changed as selective pressures change. Some, such as the courtship displays of ducks, are highly conservative and may serve as taxonomic indicators. Others, especially complex acts such as maternal care (as well as predation and fighting), are highly susceptible to change and can, but rarely do, provide evolutionary information.

One of the more bizarre episodes in the history of ethology was the attempt to demonstrate the inheritance of learned behavior, though this directly contradicts the assumptions underlying evolutionary studies. If *acquired* characters can become heritable, then what can be assumed stable?

The episode in question was the effort by the philosopher–psychologist W. MacDougal to demonstrate the inheritance of maze-learning abilities by rats. His rats had to swim through a water maze repeatedly until they eliminated all their errors. Thereupon, they were bred and their offspring, trained. After about a dozen generations, the offspring demonstrated an ability to swim the maze after far fewer trials than their ancestors had required. Most unfortunately for the hypothesis, the control offspring of untrained predecessors also showed a similar improvement.[1]

A final note needs to be added about the efforts to trace changes in learning ability as one ascends the evolutionary scale. This began with Darwin, and developed into the discipline we know today as comparative psychology. Modern comparative psychologists, of course, focus on other issues, but to describe evolutionary changes in learning ability has certainly been their cornerstone, and remains so today.

D. Mechanisms

A fly buzzes across a frog's snout. The amphibian's tongue flashes out and the fly becomes a meal. This could have been the frog's first encounter with a fly. What process or processes were responsible for its response? The answer could be very complex, indeed.

Suppose you were to design an artificial frog that would respond to flies, but not bees, and only when they are in striking range. Taking a leaf from the books of cyberneticists, we might identify the following steps: (1) detection of a change in sensory input (a fly approaches); (2) identification of the change as relevant (it is a fly, and not a bee); (3) initiation of a sensorimotor tracking program (point the snout toward the fly and follow its movement); (4) measurement of distance and decision when to fire; (5) triggering of motor response sys-

---

1. The most bizarre aspect of these experiments is that they were actually performed by an assistant, J. B. Rhine, who was simultaneously performing demonstrations of psychokinesis. As P. Medawar wryly observed, it is a bit incongruous to do, in the morning, experiments that assume no influence of the experimenter's desires on the outcome and, in the afternoon, experiments whose outcome does reflect the experimenter's wishes. But, then, parapsychologists have generally had it both ways (Medawar, 1957).

tem (tongue zips out); and (6) transformation to *status quo ante* (after swallowing, of course).

This is not a complete list. We have not, for instance, alluded to mechanisms whereby the frog's behavior changes as a result of the passing of the seasons (after a while, male frogs become more interested in female frogs than in flies), nor have we questioned that process. We have also not considered changes due to experience, or eating too many flies, or too few. In short, the question of mechanism is extraordinarily rich in research opportunities.

Mechanisms may be understood at many levels. We could, in some cases, simply describe the quantitative aspects of the stimulus–response relationship and thereby develop models that represent (or explain) the behavior. This has been the approach of the psychologists who followed the lead of B. F. Skinner. Or we could examine the neural events and their relation to activities in particular areas of the brain. A third possibility would be to provide mechanical or analogic models that describe the properties of the underlying system.

None of these approaches is mutually exclusive as none is in itself complete. In the following pages, we will give illustrative examples, i.e., notions that once had great influence. Not all, however, have survived the test of time. Many a lovely theory has fallen before an ugly fact, while others have simply been replaced by lovelier, or more encompassing, competitors.

1. DIRECTED MOVEMENTS

Some of the earliest efforts to deal with mechanisms stemmed from early nineteenth century studies of responses to light and gravity. Indeed, so successful did some of the explanations appear that a great deal of ancillary behavior, not normally considered *directed movement*, was similarly interpreted.

The doyen of this field was surely Jacques Loeb, whose ideas appeared at the close of the nineteenth century. Central to Loeb's view was that an amplified effect of a source of stimulation impinged asymmetrically upon the sense organs of a bilaterally symmetrical creature. The result of the asymmetrical stimulation was to produce a parallel asymmetry in the extent of the motor response (muscle contraction). Thus, if one eye of a toad was more stimulated than the other, the muscles on one side would contract correspondingly more than those on the other, turning the animal either toward or away from the source of the stimulus.

What has been stated for light holds true also if light is replaced by any other form of energy. Motions caused by light or other agencies appear to the layman as expressions of will and purpose on the part of the animal, whereas in reality the animal is forced to go where carried by its legs. For the conduct of animals consists of forced movements. (Loeb, 1918, p. 16)

The modern successor to Loeb's theory of forced movements is the application of control theory (cybernetics), particularly to studies of orientation and locomotion. A marvelously complete review of insect control systems, which best exemplifies the cybernetic approach, is provided by Mittelstaedt (1962).

2. INSTINCTIVE ACTS

As winter passes into spring, the Tom turkeys begin strutting, gobbling, fanning their tail feathers, courting the hens, and threatening one another. The approach of a hen produces a paroxysm of activity, which then subsides, before resuming in perhaps an hour or two. In the absence of a single hen, the Tom may be relatively more quiescent, but eventually he will perform his struts and gobbles even with the total lack of audience. What is the mechanism underlying this behavior?

The seasonal appearance of the behavior gives us one clue, and at the turn of this century and the decades that followed, many investigators followed it. The gradually lengthening days, it was shown, directly stimulated light-sensitive cells of the neurohypophysis (and perhaps nearby brain cells), leading to an increased production of particular hormones (gonadotrophins). But, beyond this, how did an increase in hormones control behavior?

The hydraulic model proposed by Konrad Lorenz provided an easily understood analogy. Imagine a reservoir slowly filling with water dripping from an input pipe. The outlet is plugged by means of a spring-loaded valve. The valve can be opened either by means of force applied to the outside of the spring or by the increasing pressure as the water accumulates. In this model, water represents the energy needed to drive a particular response (e.g., the turkey's courtship routine). The external forces releasing this energy are the stimuli emanating from the female. However, the more time since the last production of the courtship dance, the more energy (water) will have accumulated and the less force need be applied; i.e., the less attractive, willing, or nearby the female must be. Indeed, ultimately abstinence will out, and the pressure from the reservoir will spontaneously cause the valve to open and the courtship to appear *in vacuo*. The drip rate from the input is obviously hormonally controlled.

The attractiveness of this model lay not merely in its intuitive clarity but also in that it led to a number of less obvious insights into the manner in which external, eliciting stimuli and internal, including hormonal, factors interact. It held sway in ethology for several generations, especially after being incorporated by Nobelist N. Tinbergen into the hierarchical system he developed to explain behavior. It is only since the late 1960s that the utility of the Lorenzian flush-toilet model has waned to the point where its interest must now be considered largely historical.

3. CODING COLOR PREFERENCES

Gull chicks prefer pecking red to green, blue, or yellow spots. The mechanism was the subject of a study by J. P. Hailman (1967).

The retinal cells of gulls contain oil droplets that are either red or yellow and act as color filters. The yellow droplets are twice as common as the red ones. Only red light can pass through the red droplets. The cells containing them are thought to be *excitatory* in the sense that when fired, they initiate pecking. The yellow droplets, however, pass all colors of light and belong to *inhibitory* cells, those that stop peck-ing. Whether pecking occurs depends on the ratio of red to yellow cells activated. This is a simple mechanism, indeed, for an otherwise puzzling preference.

4. CONDITIONAL LEARNING

Pavlov's dog and Skinner's pigeon are so well known as hardly to bear mention, except for the sake of completeness and to allow the introduction of a caveat. The dog reflexively salivated when presented with meat. When a buzzer regularly preceded the presentation of the meat by a few seconds, the buzzer gradually acquired identical stimulus properties and soon could, by itself, elicit salivation. The response became known as *conditional* (erroneously called conditioned), as it depended on the temporal association of the primary (the meat) and the secondary (the buzzer) stimulus. In Skinner's paradigm, the pigeon was rewarded for performing an act arbi-trarily selected by the experimenter, whether pecking a key or fluffing its feathers. Again, the temporal association of stim-uli and consequences appear to assure learning.

The Pavlovian and Skinnerian models have proven im-mensely powerful tools for shaping as well as understanding behavior, yet, like all models, have some major limitations. It has become clear that the responses to the primary and condi-tional stimuli do differ in some significant, if subtle, ways. Even more important is that, contrary to Skinner, not all *dis-criminable* stimuli can become *discriminative* stimuli. For instance, rats can learn a maze-alteration habit (turn right first, left second) with food as reward, but not with water. Food and water are discriminable, but not discriminative. Rats can also be trained to avoid a certain food if the food causes nausea, even if only once, while an electric shock, even a strong one, will not discourage them from repeated ef-forts to get the food.

5. NEURAL MODELS

Octopus vulgaris will readily adjust to life in a labora-tory aquarium, a pile of bricks in one corner of the tank serving as home. From this home, it will make periodic forays to capture whatever prey organism the experimenter chooses to introduce, e.g., small crabs. By pairing crabs with one of two symbols painted on a card, and pairing an electric shock, delivered by means of a cattle probe, with the other symbol, we can teach an octopus rapidly to discriminate between symbols. One symbol will then initiate an attack, the other a retreat. Similarly, blinded animals may be trained to respond discriminatively to plastic cylinders, the surfaces of which are scored in different ways. These simple responses have served as the basis for elegant analyses of brain function.

Turning to the problem of visual discrimination first, let us consider the portions of the brain serving the visual sense. It has been shown that the optic lobes of octopus contain dendritic fields whose primary axes lie at right angles to one another. Suppose those dendritic fields represent a symmetrical grid. Then the projection of two-dimensional figures upon this grid will be represented by differences in the number of horizontal or vertical cells of the grid that are engaged. Additional experiments have shown that the horizontal projections are less easily discriminated than the vertical, a fact that accords well with the observation that in the optic lobes, vertically oriented dendrites exceed horizontally oriented ones in number. (See Fig. 1a, b.)

Tactile discriminations made by sightless animals appear to depend on the number of adjacent tactile receptors stimulated, i.e., on the proportions of grooves on the object, and not on the orientation or pattern of grooves. Thus, horizontally and vertically grooved cylinders will not be distinguished by blind octopi, provided the dimensions of the grooves are similar. Variations in the extent or width of grooving are readily detected, however, irrespective of visually detectable similarities in groove structure. In the case of the visual system, the nerve projections from the sensory endings of the arms to the brain retain a spatial organization consistent with this presumed mode of discrimination. It is hard to imagine a more striking case of isomorphism with respect to the primary sensory fields, the objects that initiated their stimulation, and the structural organization of the central nervous system itself.

Figure 1(a, b)   Octopus and pattern vision. Some patterns are more easily discriminated than others, presumably because their projections into the receptive field (the regularly arrayed neurons) form distinctly different patterns. (Photo courtesy of Bill Kier. Drawing courtesy of Ona Wang.) (*Figure continues.*)

O. *vulgaris* has also provided some evidence relevant to the question of central integrations. Wells (1959) found that in tactile discrimination tests where trials were spaced only 3–5 minutes apart, the performance of one arm showed no

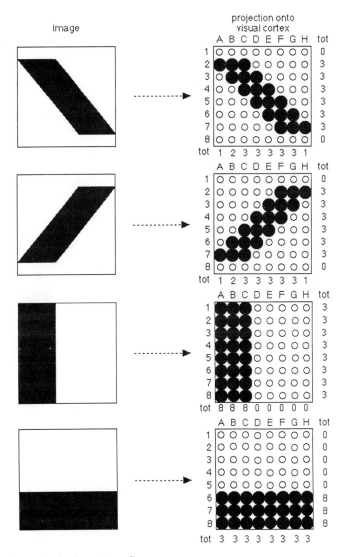

Figure 1(b) (continued).

transfer to a second arm. When the same number of trials were spaced 20 minutes apart, however, transfer from one arm to another did occur. Thus, it appears that each arm is represented by a functionally distinct neuronal field within the central nervous system. Passage of information from one of these fields to another requires an interval of time of more than 5, though less than 20, minutes.

6. HOLISTIC MODELS

Finally, we ought to allude to a family of models which, for lack of a more precise term, we call *holistic*. These are based on the finding that considerable variation is possible in the elements of a given behavior pattern, even while the pattern itself remains constant. Even so simple an act as nodding your head several times employs different muscle cells on successive nods. *Hard-wired* models can allow for a limited range of variability, but in some cases, as in the precopulatory displays of golden jackals described by I. Golani (1973), that variability simply seems unaccountable.

This is where feedback models have proven useful. Essentially, these systems examine their own output and then control or adjust the input accordingly. The best examples are complex, and given the provisional status of these models, are best omitted here. Refer to more complete sources in the bibliography. However, our premonitions and prejudices lead us to the view that holistic models will prove ever more important for understanding the mechanisms of behavior.

E. Development

It has been customary to distinguish between behavior that appears full blown (without benefit of practice or experience) and that which is gradually acquired. An example of full-blown behavior is that of tadpoles raised in anesthetic solutions; when revived, they immediately display normal swimming movements. An example of gradually acquired behavior is that of a child learning to read. The former category is often labeled *instinctive* or *innate* or *hard-wired* or *genetically fixed*. We will have a good deal more to say about this terminology in the next section. For now, we will merely assert that genes cannot control behavior, that all behavior develops as a result of a series of interactions at different levels, and that the task of the ethologist is to identify and describe these interactions.

As an illustration of how empty an explanation of *instinctive* might be, consider the gull chick that preferentially pecks at red spots. Is this instinctive? As we have described, color preferences in gulls probably result from the presence and peculiar distribution of two kinds of oil droplets in the retinal cells. How is that distribution and character determined? The simplest assumption is that at a particular locus of a chromosome (gene), a cyclical process is initiated, which leads to the production of ribonucleic acid (RNA). This RNA reacts with extrachromosomal (perhaps extranuclear) substances to produce compounds that derepress, repress, or activate other chromosomal loci. This results in the formation of another species of RNA, which also enters into extranuclear reactions. Eventually, after myriad feedback reactions, oil-soluble pigments are formed. This is one small step in the formation of a color-discriminating, pecking chick, but it makes clear how little is said by asserting that pecking at red is instinctive.

This does not, however, deny the importance of genetic differences in the development of differences in behavior. Breeds of dogs may differ significantly in temperament, performance in a variety of behavioral tests, and learning ability. But, as we will demonstrate later, the importance of genetic *differences* does not mean that genes *control* development.

Along with its interactive nature, one other important aspect of development is the age-dependency of some changes. For instance, wire-haired terriers show an increase in emotionality (according to one particular test) as they age from 17 to 51 weeks. Shetland sheep dogs (shelties), on the other hand, show decreasing emotionality during this same period. Initially, dogs of both breeds score equally in tests of applied environmental stress, but this is not the case throughout their lives; although at one age a given stress might produce the same result in both shelties and terriers, at another age the responses will differ (Scott and Fuller, 1965).

1. ILLUSTRATIVE STUDIES: BIRD SONG

The development of song in canaries, chaffinches, and other birds provides a system for following the interplay of anatomy, physiology, and experience. The attractiveness of the system lies in the precise degree to which acoustic experiences can be manipulated, by rearing in soundproof chambers, exposing to recorded sounds, or deafening surgically, and in the degree to which the resulting song can be described. Measurable visual records of the songs are detected with audio spectrographs (Fig. 2), which display frequency and amplitude changes over time. The application of this method has revealed a variety of developmental patterns.

When deafened after hatching, domestic fowl still develop their normal repertoire (although that does not mean experience is irrelevant). European blackbirds lose some of the components of normal song, but retain some phrases; and other species fail to develop normal song altogether. Many species left intact, but exposed only to recorded sound, adopt whatever songs they hear.

An example of probably the most usual situation is provided by the chaffinch, studied in some detail by W. H. Thorpe (1961) and by P. Marler (1963). The normal chaffinch song is a tripartite affair, with the third and final phrase being the most variable. An individual bird may sing several similar songs that differ in their endings. The first two phrases vary little between songs.

The division of the song into three parts, the particular notes sung, and their timing, are apparently based on what the bird hears during its first autumn. Bird song is generally under the control of testosterone. Males commence singing in spring when, under the influence of lengthening days, pituitary gonadotrophins stimulate the gonodal production of testosterone. Thereafter, the testes regress, testosterone levels drop, and song abates. In the fall, paradoxically, under the influence of shortening days, there is a brief recrudescence of the testes and a resurgence of song.

What these studies of the development of song have taught us is a far cry from the simplistic solution of the past, when song was merely labeled instinctive, as if that explained it. Though the details of song learning and production vary greatly from one species to the next, certain generalizations stand out. Songs do not just appear; their form and time of production depend on hormonal changes, the social milieu, the acoustic environment, and the season, and these factors interact in a complex and variable fashion (King and West 1988).

Studies on the responsiveness of young ducklings to their mother's call reveal other subtleties in the developmental process. Hole-nesting ducks, such as the Carolina wood duck, have long been known to attend closely to acoustic signals. The hens build their nests in the recesses of hollow trees. To emerge, the hatchlings must scale the inside of the hollow trunk, often a vertical climb of over a meter, then hurl themselves through the nest opening and plummet to the ground or water beneath. Spurs on their feet assist the climb, while their slight mass protects them from injury from the fall, which may be 10 or more meters. During this emergence of the ducklings, the mother stations herself near the base of the nest tree and utters a particular call, characterized by a frequency-modulated descending pitch.

G. Gottlieb (1970), who has spent over two decades studying the development of behavior in ducklings, synthesized a normal call, one with only ascending tones, and one which was a mixture of both normal and ascending tone. Ordinarily, naive ducklings are attracted to the first call; those reared in the absence of auditory signals show no preference between ascending and descending signals. If initially allowed normal auditory experience, and then isolated from sounds at 65 hours, no preferences appear. Exposure to a different call, the so-called distress call, leads ducklings at subsequent tests to prefer the descending call, while exposure to the distress call played backward eliminates preferences.

Gottlieb concluded that early auditory experience may induce, facilitate, or serve merely to maintain certain behavior. These three effects of early experience, he emphasizes, do not fit the traditional learning–instinct dichotomy. It is for good reason that he titles his study, "Nonobvious Effects of Early Experience."

2. EARLY ATTACHMENT

One of the most endearing scenes from the Hall of Fame of Animal Behavior is that of the gentle, bearded giant,

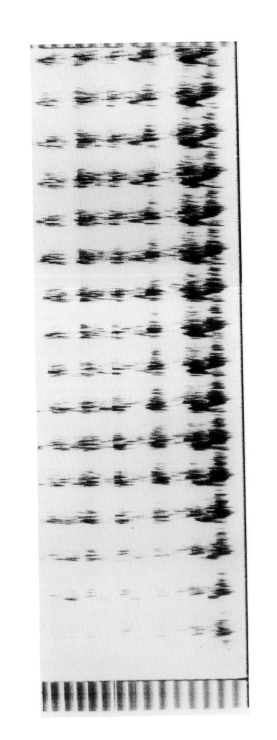

Konrad Lorenz, tramping through a Bavarian meadow with a string of goslings following him. Goslings normally follow their parents in this fashion, but Lorenz hatched these goslings in his incubator, and they adopted him as their parent. Many precocial[2] birds exhibit this behavior.

As was noted late in the nineteenth century, newly hatched chicks, ducklings, and goslings will follow almost any object moving before them. Subsequently, they continue to follow whatever first elicited the following response, to remain close to it, and even, months later at maturity, to mate with it. The process by which this occurs is known as *imprinting* from the German prägung. The appropriate metaphor is of sealing wax onto which a pattern can be impressed, but only during the brief period when the wax is soft. Once it has hardened, the pattern imprinted upon it is permanent.

Lorenz (1952) believed the same applied to his goslings; if exposed to a moving man 12—20 hours after hatching, their *critical period*, goslings would develop a lifelong preference for men over geese. We now believe that the process is not as irreversible as all that, nor is the critical period as sharply bounded as Lorenz once felt. The general features of imprinting are recognized to accord with what Lorenz originally described.

Imprinting makes clear to ethologists the mechanism that assures certain behavior. Precocial animals, for instance, can readily wander away from their nest or parents, and some mechanism must exist to assure they do not do so. Imprinting is one such mechanism, and it appears to be used not only at birth. For songbirds, physically restricted to their nests as naked nestlings and therefore allowed to learn the character-

---

2. Precocial refers to animals that are able to locomote and feed independently when in contrast, altricial species, such as rabbits and robins, are helpless when first born and confined to a nest.

Figure 2   Audio-spectrograph. In these groups of bird-song, time is displayed on the horizontal axis, frequency (pitch) on the vertical axis, and intensity (loudness) by the darkness of the lines. (Photo courtesy of Steven Nowicki.)

istics of their parents more gradually, imprinting may be important later, for learning the details of their habitat, or the characteristics of their species' song.

The features of imprinting that set it somewhat apart from other types of learning include the rapidity with which it occurs (a 10-minute following session may suffice to fix a gosling's preference), its stability over time, the absence of any obvious reward such as food, and the fact that the significance or consequence of the established preference may not emerge for many months (as when sexual preferences are formed for the object initially followed). Actually, these are not altogether distinctive attributes; imprinting can no longer be regarded as fundamentally different from other sorts of learning, yet, it is qualitatively distinctive enough to deserve special attention.

The basic notion in imprinting is that the organism is a blank slate on which a preference can, at a certain time, be recorded. A number of investigators, however, have now shown that only certain preferences can be recorded, and that some are far more easily or permanently recorded than others. A strain of deer mice, *Peromyscus sp.*, whose normal habitat is fields, was reared for several generations in the lab by Wecker (1963) and then allowed to raise young in fields or woods. Thereafter, the young's preferences for woods versus fields were examined. A short exposure to fields did produce a corresponding preference (habitat imprinting), but exposure to woods led to neither habitat being preferred. Apparently, deer mice can be imprinted onto fields, but not woods.

What about neurophysiological underpinnings of imprinting? Because it takes place rapidly and at a particular time in life, one might expect it to correlate with specific changes in the central nervous system. In fact, certain changes in the rate of protein synthesis at specific cerebral loci have been noted by Horn *et al.* (1973), but the general significance of these findings for a neural theory of learning has yet to be found.

The nature of imprinting (and presumably all learning) appears to be interactive. Some clues that reveal this can be gleaned from a study of imprinting in goats. Among goats, it is not the kids that imprint onto the mother, but the mother to her kids. In a series of studies extending over several years, several investigators established that allowing a goat that had just given birth 5 minutes' contact with her kid will generally assure the kid's subsequent acceptance, even after a 3-hour separation (Klopfer, 1971). Denied the 5-minutes, however, a separation after birth for as little as an hour will likely lead to permanent rejection of the kid. If several kids are born, but only one is presented for the required 5 minutes, the others are nonetheless subsequently accepted. Alien kids, though, are not accepted. However, if the goat is rendered temporarily anosmic (without the sense of smell) at the time of birth, and then given 5 minutes with one kid, she will subsequently accept both her own *and* alien kids. If the anosmia is induced only when the kids are presented 3 hours after birth, the kids, whether her own or aliens, are as often accepted as rejected. Finally, when an alien kid is presented during the critical 5 minutes after birth, it is subsequently accepted, and so are all the goat's own kids, which she has not contacted. Other aliens are, of course, rejected. (See Fig. 3.)

Somehow, parturition makes a goat maternal. However, her maternal responses are fixed onto specific animals as a consequence of giving her a brief exposure to them. All her own kids are treated as alike, but alien kids are treated as individually distinctive. Because this discrimination is lost under anosmia, olfactory cues must play a role; as the onset of responsiveness is triggered by parturition, internal changes are clearly important, too.

These internal changes associated with parturition include the release of oxytocin from cells in the midbrain, though manual dilation of the cervix will accomplish the same result. If a balloon is placed into a virgin sheep's cervix and inflated several times, so as to simulate birthing, oxytocin appears in the blood (Peters *et al.*, 1965). This presumably is also true for goats. On the basis of these and a few other findings, Klopfer (1971) proposed this sequence of events to explain mother–young bonding in goats. The presentation of the head during birth follows a cervical dilation that produces a release of oxytocin. Oxytocin's residence time in the blood is very brief, but during that time the animal's olfactory sensitivity is altered, allowing for olfactory imprinting. The oxytocin

Figure 3   Mother and young. A newborn kid is sniffed and licked by its mother.

also directly alters the behavior of the animal by acting upon neurons in the brain. Thus, the few minutes after parturition transform the doe into a mother ready and eager to attach herself to a kid. Once attached, she displays distress on separation. Spared attachment, the absence of a kid leaves her as nonchalant as any virgin, despite a distended and dripping udder.

3. LEARNING TO FEED

Because feeding is an essential prerequisite to survival, it is scarcely surprising that this response, too, has attracted considerable attention from those interested in development. That interest has most recently been heightened by the discovery by Hall and Williams (1983) that, at least in rodents, suckling and later feeding are not homologous behavior patterns. These seemingly similar responses apparently develop and are controlled quite independently. The findings are still too recent and startling to have been applied to other species, but we must keep them in mind as we review older work on the development of feeding.

In precocial birds such as domestic chickens, the young must feed themselves, and they do. However, feeding efficiency is initially low, and increases gradually. A chick on its own takes much longer to peck at objects than it would in a group, because chicks imitate one another, and with more chicks the probability of any one pecking in a given amount of time is increased. The presence of a hen facilitates pecking even more, and helps direct the chick toward appropriate objects from the start. In her absence, pecking is often directed at bill tips and toes, and only accidentally does the chick discover an edible particle of appropriate size. The accuracy of that pecking increases with practice. By recording pecks on slabs of soft plasticine, E. Hess (1956) showed that pecks were widely scattered, but in time came to be properly concentrated. When the chicks were made to wear prisms that displaced objects to one side, the same improvement occurred, even though the chicks never received the food that was the object of their efforts.

The chicks of some gulls start out in much the same way, pecking particularly at the contrastingly colored bill tip of the parent. As the parent then regurgitates food from its crop, the chick comes into contact with it and learns its taste. Soon, it recognizes the regurgitated mass and will peck at it when it lies on the ground and thus comes to attack food directly, rather than waiting for it to appear between the parental mandibles (Hailman, 1967). (See Fig. 4.)

Aspects of feeding need to be developed among adults responsible for feeding the young, too. Pigeons feed their squabs by regurgitating a material (crop milk) secreted within their crop. The initiation of the secretion is under hormonal control. Hormone production, in turn, is stimulated by exposure

Figure 4   Gull chicks learning to feed. (Photo courtesy of Jack and Liz Hailman.)

to eggs and squabs, according to D. Lehrman (1965). It was thought that the uncomfortable sensations associated with an engorged crop served to start feeding. Presumably, a squab sticking a bill inside the parental mass triggered a vomiting reflex, which in turn brought about depletion of the engorged crop, and relief to the parents. There are some questions about the validity of this interpretation. Naive parents still feed their chicks when the crop has been anesthetized and they presumably feel no pain, but the interpretation does suggest how internal and experiential factors may interact in development.

We conclude this section with a human example that illustrates not only interactions, but also the rapidity with which they may occur. Some women may, on parturition, have breasts of such a size or shape as to make nursing by their infants difficult. Specifically, the first time the infant attempts to suck, its upper lip may be forced back upon its nostrils, blocking inspiration. The baby, of course, gasps, stiffens, and pulls back, in the process seeming to push its mother away. The inexperienced mother's behavior is immediately and perceptibly altered by this rejection, and the next effort to induce nursing is far less spontaneous. The infant's inevitable response is thus doubly effective in disrupting the relationship. An experienced mother (or assistant) who, anticipating the problems, pulls back the lip from the occluded nostril, can save the day (M. Gunther, as reported by M. Mead, personal communication, 1964). Similar problems often arise among domestic animals, particularly sheep and goats with oversized teats. It does not take long for a female with a tender udder to tire of a persistent youngster whose futile nursing efforts are a literal pain. There is more to nursing than simply sucking, and in the case of humans, it often takes experience or assistance for this to be learned.

4. METAPHORS FOR DEVELOPMENT

The application of a template to temporarily softened sealing wax, which, on hardening, holds the pattern, is an attractive metaphor for imprinting. But other metaphors are also exploited to explain or describe development, as in metamorphosis, where a crawling caterpillar is transformed into the volant moth (or butterfly), or the adage, "As the twig is bent, so grows the child."

Although the cases of behavioral metamorphosis have not been studied quite so much as imprinting in birds and mammals, many are well known. For instance, the male puppy that squats in the manner of its sisters when it urinates will, at a particular age, spontaneously lift a rear leg. This behavior does not seem to require any particular experience; as long as the nervous and endocrine systems are allowed normal maturation, the behavior metamorphoses at the appropriate time.

The bent twig paradigm implies that the continuity of influences produces the final outcome. In sum, we have three different metaphors; which applies when? Is early experience as important as all three of these metaphors imply?

Our review of some of the research in this area prompted the following conclusion.

The most significant feature of development is its ability to respond to perturbations. This would seem a truism, except that its implications are often overlooked. Creatures whose environments are full of surprises would be ill served by developmental rules that were so rigid as to be unresponsive to changes. Species with relatively low reproductive rates and long life spans, such as primates, require alternative pathways that allow development to continue in the face of perturbations ... because compensatory processes are also brought into play.

An illustration of how compensation may operate was provided by a study with impaired kids (Klopfer and Klopfer, 1977). Goats generally produce twins yet ... individual differences in temperament are evident. We were interested in whether such differences could be traced to differences in the treatment the kids received from their mother early in their lives.

Upon their birth, the heavier of two kids is initially the more vigorous and thus receives more maternal attention. This attention, however, usually given in the form of licking, often knocks the kid off its feet and thus slows its progress to the teat. The result is that the initially weaker kid often nurses first, and early differences in vigor are diminished... .

The parent is, of course, but one element in the environment of the young, and might be expected to be the most important factor that regulates events and mitigates potentially calamitous challenges. But, even physical factors beyond the parent may act as buffers, just as (in ducks) heightened humidity, to an extent, can compensate for slightly cooler temperatures during incubation. An appropriate analogy can be taken from ecologists, whose descriptions of complex food webs, which are hierarchically ordered levels of interactions, include specification of causal links within and between levels. What is striking about these foods webs is that there usually exists more than one path for energy flow from any one producer to any one consumer. Hence, a perturbation at one point—over- or underproduction due to a short-term environmental shift—is not necessarily propagated, but may be buffered and its effects dissipated throughout the system, and these adjustments may not be predictable. Only in the simplest webs, with no alternate pathways, as in some arctic habitats, does a change at one point invariably affect all others (Klopfer, 1973). Similarly, it may be solitary species whose life stories do not involve changes in the consort pool that show most strikingly the effects of particular early experiences. The more social the organism, and the more varied its social network, the less likely that any one experience will dictate an outcome (Klopfer, 1988).

5. FINALLY, THE GENE

We cannot leave the topic of development without a word about genes. Only their DNA is literally transmitted from one generation to the next. Hence, if behavior seems also to be inherited, it is reasonable to assume the behavior issues from the gene. Only after careful analysis do we see why this cannot be so, not that this means the environment is the controlling factor. Both parties to the venerable nature–nurture dispute ("It is due to the genes" versus "No, to experience") are equally in error (Lewontin *et al.*, 1984).

Consider first what we mean by behavior. Take a specific stereotyped act such as the gull chick's instinctive pecking of its parent's bill. It has a stereotyped and unique character; each repetition resembles its predecessor and is distinctive from other kinds of acts. Yet, slow-motion analysis reveals hidden complexities. This seemingly unitary response is really composed of several discrete movements; a step forward, a twist of the neck, and a lunge with the head can be separately distinguished. And these components, too, are susceptible to a finer analysis. No two neck-turns are precisely identical, and even if they were, different muscle fibers would cause them. The overall peck may seem to be the same act, but in fact each peck is distinctive. What exactly is contained in or coded by the genes?

In earlier times, it was believed that a miniature person, a homunculus, was contained in the head of the sperm. Development consisted merely of its growth within the nutrient medium of the womb. The homunculus obviously had an array of even smaller homunculi in its body, the entire lineage of humankind. With the discovery of cell fertilization and the improvement of optical microscopes, the homunculus disappeared, at least in its most primitive form. But when we claim that a gene stands for pecking, we are really invoking much the same concept. Genes, portions of the double-stranded chromosomal DNA, through their interaction with their environment, can specify the character of particular polypeptides or proteins. These in turn may interact with the chromosomal material and other extrachromosomal (environmental) material to produce other molecules. How these primary transcription products ultimately combine to further the course of development is one of the mysteries still far from resolution. Yet, it is clear that it is through *interaction* that the organisms develop, not through the reading of a genetic blueprint, and not through shaping by environmental forces. Behavior is like a drumbeat, which is made neither by the drumsticks nor by the drumhead, but by their interaction.

While acts may be universal or unique, variable or stereotyped, reversible or irreversible, learned as a result of prolonged training or expressed properly on the first try, these distinctions do not pair symmetrically. All humans speak, but they speak different languages. They are born with tongues, but as a result of an (nonheritable) accident, a tongue may be irreversibly lost. Some experiences have lifelong effects, while behavior shown at birth may later be lost. Thus, it is a mistake to assume that highly variable behavior is for that reason the result of different processes from rigidly fixed behavior, the one the result of experience, the other, of genes.

Differences in behavior are, of course, heritable. That is what accounts for breed differences. But this statement literally means that if two individuals differ in a genetic, heritable fashion, then the interaction between their genes and their environment may also differ in a manner that produces behavioral differences. However, the interaction products could conceivably be the same, so we must not conclude that the behavior is genetic.

Behavior is a phenotype, an observable characteristic. Genotypes, not phenotypes, are inherited. A crucial issue in the study of genetics is the relation between genotype and phenotype. What nongeneticists (and some geneticists) may fail to realize is that the postulated relationship is based on statistical procedures. In a given population, we observe a range of phenotypes, V_P, associated with a range of genotypes, V_G, a range of environments, V_E, and the gene–environment interaction, V_{EG}. We may assume that $V_P = V_G + V_E + V_{EG}$. V_P is always greater than zero, for no two individuals, even identical twins, are exactly alike. We can then compare two populations of organisms that are raised in identical environments ($V_{E1} = V_{E2}$) and that differ in only one gene. Suppose their phenotypes (behavior or eye color) differ. Is it due to that gene? We cannot say, because we cannot independently evaluate the significance of the interaction term V_{EG}. All we can conclude is that the behavioral difference is related to the gene difference. This says a lot, of course, and for the animal breeder it will suffice. It does not, however, allow the student of behavior mechanisms to conclude that the gene in question contains, causes, or codes for the behavior. Organisms, as they develop, interact with the environment, thereby generating the information by which they are described. Clearly, development is not an unfolding of imminent characteristics, or the growth of a homunculus.

F. Suggested Reading

Almost any of the popular texts on animal behavior does justice to these topics. The examples used here are elaborated in *An Introduction to Animal Behavior: Ethology's First Century*, by P. H. Klopfer (Englewood Cliffs, New Jersey: Prentice-Hall, 1974). R. Fagen in his *Animal Play Behavior* (Oxford, England: Oxford, 1981) has done a marvelous job developing many of the issues we raise in the context of play. The difficult and controversial problems of development and inheritance of behavior are incisively addressed by R. Lewontin, S. Rose, and L. Kamin in *Not in Our Genes* (New York, New York: Penguin, 1984), S. Oyama in *The Ontogeny of Information* (New York, New York: Cambridge, 1985), and J. Klama, *Aggression: The Myth of the Beast Within* (New York, New York: Wiley, 1988).

3

Prospective Problems

Not all of the issues of adaptation, evolution, mechanism, and development have been resolved. They remain *prospective* to much the same degree that they were the focus of attention in the last several decades. Yet, the general principles that guide us in searching for solutions to particular problems seem to be well established. Established principles can always be cast aside, of course, and they also have a disconcerting way of dissolving when least expected. But, for the moment, the principles annunciated in the foregoing pages seem secure. This is not the case with the issues that we group under the heading prospective. Here we deal with issues which, while they have a venerable history—for instance, studies of animal orientation go back to the last century, and earlier— are in too great a flux just now to allow us to be at all certain about what the future will bring.

Consider, first, the status of studies of animal cycles, navigation, and orientation.

A. Directional Movements and Rhythms

Even before Darwin's studies on the subject (*The Movements and Habits of Climbing Plants*, 1876), publications on directed movements of both protozoans and plants had become available. In addition, descriptions of geotropisms, or movements in response to gravitational forces, and phototropism, or similar responses to light, had appeared in the scientific press. Throughout the latter half of the nineteenth century, various authors described other directed movements on the parts of plants, protozoa, and bacteria in response to chemicals, moisture, pressure, temperature, and other stimuli. The most important synthesis of the ideas advanced in these studies was produced by Jacques Loeb in 1890 and 1906 (compare Klopfer, 1974). In brief, Loeb believed that whenever a source of stimulation impinged asymmetrically upon the sense organs of a bilaterally symmetrical creature, a similar inequality in the magnitude of muscle contraction on the opposite sides of the animal would result. The effect would be to alter continually the position of the animal until its sense organs become equally (i.e., symmetrically) stimulated and the bilateral muscle groups equally active. This was considered an accidental process and of little, if any, adaptive value.

The second element of Loeb's theory concerned the mechanism of the orientation process. It is here that stress was placed on the symmetry of receptor and effector activity. Here too, the shortcomings and rather restricted applicability of what was intended as a universal explanation become apparent. This was shown in experiments in which orientation was not impaired either by the removal of a limb from one side or by other artificial interference with normal body symmetry.

The modern successor to such mechanistic approaches is seen in the application of control theory (cybernetics) to directional orientation. Consider the orientation of a spider to a potential prey organism. When one of its four lateral eyes detects an object, it turns toward it, just the right amount, so that the object will fall into the field of the frontal eyes. The amount of the turn is specified in advance and is independent of visual feedback. Is the command to turn based on a specific turning time? Apparently not, because the speed of turning may vary. Nor is the proprioceptive input (information from position sensors) from the legs necessary. However, since the angle through which the animal turns with each step is constant, the turning command may involve the computation and specification of a discrete number of steps.

In addition to displaying directional responses, many organisms are also cyclically active. "Rhythmicity is a characteristic of nature" (Cloudsley-Thompson, 1961), whether we consider sidereal motions, the alternation of the seasons, or the pumping of the heart. Scientific attention was first focused on animal rhythms at the beginning of this century. Von Buttel-Reepen and Forel (Klopfer, 1974), observing the regularity with which bees sought food at Forel's breakfast table, concluded that bees possessed a time sense. This, in turn, implied (though it did not prove) the existence of a stable interval cycle or rhythm, and this was probably the first rhythm to be seriously studied. Such rhythms, in turn, are probably fundamental to cyclically recurring behavior such as diurnal activity cycles, as well as to the complex pattern of responses involved in animal migration and navigation.

The study of biological rhythms, their functions and controlling mechanisms, is an absorbing branch of biology. By contrasting the various ways in which cyclic biological phenomena have been catalogued, we can then distinguish four different schemes of classification. The first of these describes biological rhythms or cycles in terms of their frequency (i.e., whether annual, daily, or hourly). The second is concerned with the overt expression of the rhythm (i.e., whether it is a change in a specific metabolic function or in gross behavior). The third depends on the adaptive function of the rhythm. The fourth, finally, is based on the controlling stimuli; in other words, it centers on the nature of the basic timing mechanism. This last, of course, involves ethologists in cellular physiology and biochemistry.

We can distinguish cycles that are more or less than 24-hours. The latter include 2-hour cycles of activity or feeding in laboratory rats and other rodents, and 4–5-hour cycles in human infants. These cycles are linked with periodic changes in gastric motility and hunger. In a salamander marked change in thresholds for tactile stimulation occur at 8–10-hour intervals. Rhythms such as these are polycyclic, a term including any function that has a period of 12 or fewer hours.

Diurnal periods, in contrast to those that are polycyclic, occur but once in 24 hours. The term diurnal, however, is misleading in this context, because many animals with 24-hour activity cycles are maximally active at night. Properly speaking, theirs are nocturnal periods. Further, few 24-hour cycles are held precisely at 24 hours. They usually range from 23 to 25 hours, though any one individual may show an extraordinary constancy in his own period. Hence the term diurnal period has been replaced by circadian rhythm (*circa*, about; *dies*, a day). Circadian rhythms appear to exist for almost every animal or function in which they have been sought.

Periods of more than 24 hours include estrous and sexual activity cycles, which vary from 4-1/2 days (laboratory rats) to 12 months (in seasonally breeding birds, mammals, fish, and insects). Also included here must be the 6-day cycles in the feeding of some insects and in the phosphate content of rabbit muscle.

Many allegedly cyclic phenomena represent statistical artifacts. Nowhere is this point more cogently and humorously made than in an article by Lamont Cole (1957). His "study" demonstrated the existence of a rhythmic change in the metabolic rate of a unicorn. (Because unicorns are mythical creatures, he had to be very ingenious to obtain metabolic data. What he did was to turn to a table of random numbers and let successive numbers represent an increase or decrease in metabolic rate, depending on whether they were odd or even.) The fact that the mathematical techniques used in this demonstration are similar to those often used by workers in the field of chronometry should induce a healthy measure of skepticism toward many reported "cycles." Consider such claims as that eminent people tend to be born in greatest number in winter or that political revolutions follow the sunspot cycles.

The adaptive features of many cycles have yet to be discovered. "What is the significance of cycles in body temperature, for instance?" "Is it possible that some cycles may represent incidental by-products of other processes and are in themselves of neither value nor harm, analogous to the secondary (Pleiotropic) effects of certain genes?"

Annual cycles of reproduction, timed by environmental agents of widespread or universal occurrence, assure simultaneity in male and female sexual behavior. In most temperate-zone birds and some mammals, the external timer for this reproductive rhythm is a particular day length. Thus, all members of a particular species residing at a particular latitude will show phase synchrony in reproductive processes. Diurnal or polycyclic rhythms of reproduction have the same effect.

Daily activity cycles may aid an animal to avoid predators. (Of course, a nocturnal animal's predator may also evolve efficient means for nocturnal hunting; in this case the benefit is the predator's.) In areas where ducks are hunted daily except Sunday, a 7-day cycle of tameness has been noted. Nocturnal or cyclic activity may also provide fewer competitive feeding opportunities, especially if the cycles of the most similar sympatric (occupying the same range) species are out of phase with one another.

Protection from extremes of temperature or water loss is essential to many desert-dwelling forms. This protection is often afforded by a subterranean habit, emergence being timed to coincide with the setting of the sun. In animals of the seashore, cyclic burrowing or emergence is probably an adaptation that prevents desiccation, or, in other cases, inundation.

Finally, rhythms are important in the determination of direction during migration, allowing the possibility of correcting for the apparent movement of the sun, as well as in the timing of migration.

Many cycles quickly disappear when constant environmental conditions are imposed on their possessor. Thus, changes in the heart rate of a frog, which may show a regular diurnal alteration, will cease when the animal is confined in a constant-temperature chamber. The other extreme is represented by cycles that persist for a considerable time even after isolation from all pertinent and cyclic environmental stimuli, such as the activity rhythms of many rodents. An intermediate situation is afforded by functions requiring an environmental cue for their initiation or timing, but which then run independent of environmental cues. Many different schemes of classification and nomenclature have been promoted to deal with these different kinds of cycles.

The distinctions between the categories listed are far from absolute. Animals cannot simply be classified as to the nature of their rhythms, because one animal may have many cycles, differing in function, phase, frequency, and degree of dependence on environmental timers.

When the cycles are driven by external events, the analysis of their control mechanisms generally poses little problem. One of the earliest cases, Rowan's (1925) investigations of the seasonally cyclic recurrence of reproductive behavior in certain temperate-zone birds and mammals, showed photostimulation of a certain magnitude to be the relevant timer, e.g., when days attain a certain length, reproductive activities commence. Recent work has elucidated the pathways along which the reactions initiated by photostimulation must travel, as well as the fact that factors other than photostimulation may be involved. For many vertebrates the primary pathway is photosensitive cells in the brain or in the rudiments of the pineal gland. This demonstrates that the seasonal variation in day length can drive the reproductive cycle. Under constant conditions the rhythm may break down.

When the cycles are largely endogenous (i.e., independent of environmental cues), the situation can become more complex. The attempts to explain the persistence of rhythms in animals held under seemingly constant conditions have fallen into two categories. On one hand, one assumes that truly constant conditions are unattainable, that some residual and cyclic variables always remain to impinge upon the isolate, e.g., changes in cosmic radiation. On the other hand, we have the assumption that physical or chemical reactions at a subcellular level may oscillate independent of environmental factors and serve as the basic driving mechanism for a biological clock. Such self-sustaining oscillators may be entrained by environmental rhythms whose frequencies do not differ too radically from the endogenous frequencies. This would account for the phase-constancy of most biological rhythms, while the same rhythms, under constant conditions, drift out of phase with environmental rhythms.

B. Migration, Orientation, and Navigation

Studies of seasonal movements of animals must be concerned with two distinct aspects, the causative factors *per se* and the processes that allow the maintenance of a particular course. Analysis, in turn, must consider two kinds of causes, distinguished by Lack (1954) (and Aristotle) as proximate and ultimate. By *ultimate* is meant the historical or selective factors favoring the evolution of migratory behavior. Those most commonly cited include climatic changes necessitating annual movements, continental drift, and adaptations for the exploitation of a temporary or fugitive food supply. In short, the animals that did not migrate, starved, while those that moved, survived. *Proximate* factors are those that immediately trigger the migratory response. These were alluded to in the previous section where the role of photostimulation in the induction of breeding behavior was considered. Most of the same comments apply to the initiation of migration.

The maintenance of a particular direction represents the second distinct aspect of the problem, one which, as we will see, involves an intimate relation between migrations, tropisms, and rhythms. The earliest studies were greatly limited by the inability of investigators to keep track of specific individuals. The actual goals and pathways of migrants could only be indirectly inferred. Marking of individual birds for tracing had been proposed at the start of the nineteenth century by J. A. Naumann (cited by Stresemann, 1951) and again by Borggreve in 1884. Not until 1890 did such a scheme become adopted, however; in that year a Danish school teacher, H. C. Mortensen, introduced leg rings. Mortensen's bird bands were quickly adopted by ornithologists the world over, and clubs were developed to store and exchange information on recaptures. Thus, a fairly complete picture of migratory pathways of birds was painted, opening the possibility for the experimental study of orienting mechanisms.

One of the earliest and, for its time, most sophisticated of these studies was that by Exner (1893). In order to determine whether homing pigeons receive information on the direction of displacement while being transported from home, he moved several birds while they were under anesthesia. Others were transported in swinging or rotating cages (presumably to eliminate inertial clues), while yet others were subject to the discharge from a faradic cell (to interfere with electromagnetic clues). Regrettably, Exner's sample was too small, and the response variance too great to allow any conclusions to be drawn. Surprisingly, it took over 50 years for biologists to repeat some of his important experiments.

However, other results suggested that orientation depended either on an inertial sense or on the perception of electromagnetic fields. Cases were cited in which homing pigeons were delayed by electrical storms, and reports collected that birds in the vicinity of radio stations were disoriented. More recently, similar reports of disorientation as the consequence of radar beams have been made. Claims that interference with the electromagnetic sense of pigeons (by attaching magnets to them) prevented orientation were not at first substantiated, though often repeated. Until 1970, it was believed unlikely that there were physiological sensitivities of low enough threshold to allow detection of changes in electromagnetic fields. That has now changed.

Exner's other suggestion that orientation involved retracing the directions followed on the outward journey was taken up by others. However, we now know that retracement is not always involved, as seen by the direct observations of homing birds.

Not surprisingly, another of the early explanations of homeward orientation involved the assumption that the goal (home) itself provides a beacon that can be seen or otherwise sensed. The arguments by Watson and Lashley (1915) convinced most biologists that in birds, visual mechanisms were simply insufficient to account for long-distance homing, e.g., because of the earth's curvature, light from the top of a 500-mile distant lighthouse would be only visible from an altitude of 25 miles.

In 1944, Donald Griffin, the discoverer of bat sonar, listed the major theories accounting for orientation and the work that supported them. These theories involved explanations based upon either vision, kinesthesia, or electromagnetic receptors. Griffin concluded that none of the experimental data favored these theories and proposed instead orientation based on recognition of major geographical or topographical features, typically prevailing air masses, and the relationship between such features of localities near their home and the homeward direction. Griffin also suggested that celestial clues such as the direction of sunset relative to home might be used. This was portentous. Only a few years later, Kramer and his students were able to demonstrate the existence of a sun compass in certain birds.

Another classification of theories was offered by Matthews in 1955. Matthew's first category included theories that required the maintenance of sensory contact with home. Theories in this category (and the second, below) were rendered untenable by Griffin's (1944) criticisms. The second category included theories of navigation "by means of a `grid' derived from the earth's rotation and magnetism" (Matthews, 1955), and the third, a grid derived from the sun's coordinates. To this last, we would today have to add coordinates of other celestial bodies. Theories in this last class have proven most viable. True navigation may involve a sun compass and chronometer, but in addition, certain other components are needed. These latter must include either mechanisms for recording the direction and distance of the displacement from home or an almanac that allows the animal to compare the coordinate of its home with its displaced position.

An interesting sidelight that related the clockwork to the neuroendocrine system was revealed in a study by S. Emlen (1969). He found that a particular star pattern (simulated in a planetarium) could trigger directional movements either southward or northward, depending on whether the birds had previously been subjected to a long-day (springtime) or a short-day (autumn) regimen.

Any number of inertial navigation systems could meet the necessary requirements for avian navigation. Some systems track changes in acceleration and direction on the outward journey and integrate this information in order to plot the return. Barlow (1964) has detailed the characteristics of some inertial devices and discussed the possibility of there existing biological analogs. Thus far, however, there is little experimental evidence supporting the existence of inertial theories. The possibility that inertial and visual mechanisms coexist in one animal and can replace one another (hybrid systems) depending on environmental conditions makes the design of critical experiments exceedingly complex.

As for visual orientation systems, these all appear to depend on the existence of a grid. One of the two necessary coordinates of such a grid appears to be based upon the constancy of the sun azimuth at a particular locality, date, and time. Knowing what the sun azimuth should be at noon on a particular day at home, for example, we can readily determine whether we are actually at home or have been displaced to the east or west. In theory, information on displacement to the south or north should be provided by comparable data on sun altitude. Modifications of this have also been offered, grids based upon values for local rates of change in altitude and azimuth. In laboratory and field experiments, however, the altitude of both an artificial and the real sun have not been seen to influence directional choices of birds, although fish may make some use of information on solar altitude.

Unfortunately for aspiring theorists, a number of other seemingly contradictory results have yet to be reconciled. While, on the one hand, Kramer and his students had repeatedly noted disorientation of birds tested under heavily clouded skies, i.e., with no sun visible, they also reported orientation in the absence of all celestial clues (Schmidt-Koenig, 1979). Old World warblers were tested in a circular cage, fitted with perches about its periphery that automatically recorded the arrival or departure of a bird. During the spring and fall, captive migrants frequently became very restless, flying to and fro and orienting themselves along their usual migration track. The birds maintained this orientation both inside an opaque shelter and in the confines of a climatic chamber. Only when the steel door of this chamber was closed did the directions chosen by the birds appear to scatter randomly. It is possible that this and similar results are artifacts of small samples and inadequate or inappropriate statistical treatment. However, at present these results remain somewhat disconcerting to proponents of exclusively celestial theories.

One other experiment must be mentioned for the support it gives to proponents of inertial theories. Migratory pigeons, after having been rotated, show a peculiar pattern of electrical activity in the cerebellum. Nonmigratory races fail to show the same pattern. It is because the cerebellum is so intimately involved with balance and coordination that these results are of particular interest and importance.

Since 1970, orientation studies have increasingly focused on the nonvisual modalities. Animals other than birds, such as arthropods, marine turtles, and amphibians have also been studied, but most often the effort continues to be focused on the avian navigational system, and particularly that of pigeons.

In the 1960s, several studies reopened the question of magnetic sensitivity. Yeagley's (1947) findings of interference effects from magnets had previously been rejected by most biologists, so these newer studies were initially treated skeptically. By 1971, when Keeton asserted, "Magnets interfere with pigeon homing," the skepticism began to be replaced with new experiments. Experiments in which pigeons were fitted with frosted semiopaque contact lenses, but still homed, also indicated that the eyes, while normally used, need not play an essential role in homing. Further experiments utilizing birds wearing magnetically opaque hoods or birds trained to discriminate magnetic fields in a Skinner box are underway now. An assessment of the information available on the earth's magnetic field supports the optimism with which these studies are being pursued, as does a theoretical analysis by Kiepenhauer (1984), which makes explicit how magnetic fields could provide an orientational grid.

A final set of proposals has excited much attention since the 1980s, namely, that pigeons home using olfactory cues. Experiments in Italy with Italian pigeons seem to substantiate this notion, though replications elsewhere do not. Are Italian birds more refined sniffers? The geographical shape and position of Italy might make olfaction a more useful modality than would be true in central Europe or the United States. Still, the limited olfactory sensitivity of birds in general and pigeons in particular, not to say the absence of compelling experimental evidence, make olfactory navigation a less-than-robust prospect. Schmidt-Koenig and Ganzhorn (1990), reviewing the often hostile polemics that have characterized the field of orientation, suggest that birds may, in fact, be able to navigate using many different kinds of cues. Those any particular bird comes to depend upon (visual, electromagnetic, olfactory) may depend on the bird's early experiences. In an area where odors don't map in a consistent fashion, presumably pigeons would not try to follow their noses.

C. Hormones and Behavior

1. WHAT ARE HORMONES?

In an age where every schoolgirl knows about "raging hormones," it may come as a surprise that the very concept of hormones is of relatively recent origin. The demonstration that these are secretions directly into the blood stream by glands lacking ducts, dates back no farther than the 1920s, when cells of the pancreas, the islets of Langerhans, were shown to secrete insulin. As is usual, Aristotle and his followers had anticipated modern endocrinology (the study of hormones), but their "humors" were ill defined and postulated on logical grounds rather than experimentally demonstrated. The generally accepted evidence of the presence and action of a hormone rested upon the demonstration of a specific effect on its being withdrawn, and a reversal of that effect on its reintroduction. Thus, in the absence of the islets of Langerhans, blood-sugar levels rise and sugar is excreted in the urine, a symptom of diabetes. If an extract of the islets is injected into the bloodstream, there is an almost immediate drop in bloodsugar. The extract is known as insulin.

Hormones are defined as specific chemicals released into the bloodstream, usually in minute amounts, by ductless glands, or groups of aggregated cells. Some of these cells are imbedded within other glands, as are the insulin-secreting cells within the pancreas; others may lie within nervous tissue, for example the oxytocin-secreting cells in the midbrain. (Oxytocin is involved in uterine contractions and, in some species, the development of maternal behavior.) Some hormones act upon specific targets, as insulin influences cells of the liver to take up or release a sugar. Others may influence several different organ systems; adrenalin, for instance, is involved both in changes in muscular activity and in emotions. Some are similar across species. Mammalian sex hormones will exert similar effects on mammals, birds, or frogs. Others, however, are quite specific in their actions. Oxytocin may lead to maternal behavior in sheep, goats, or rats, but there is no evidence of a similar effect in primates, dogs, or cats.

Many hormones have several different effects, or one gland (viz. the pituitary) may secrete several hormones. Some hormones have regulating effects, maintaining constant levels of sodium, sugar, or water. Others stimulate the onset of growth of certain organs, as the gonadotrophins from the pituitary gland stimulate the growth and development of testis and ovary. They may trigger specific responses, such as ovulation, milk let-down, or migration. Finally, some hormones control the production and release of other hormones. Clearly, endocrinology is a complex field, challenging scientists in a host of different disciplines. Nowhere is its complexity greater than in the realm of animal behavior, nor have we made more than the merest start in understanding the subject. We will try here to illustrate how hormones act by tracing the development of our understanding of the relations between hormones and sexual behavior.

After E. H. Starling introduced the term hormone in 1905, three steps were needed to study hormonal effects on behavior. First, a correlation was sought between the size and activity of a presumed gland. Thus, birds sing, migrate or mate only after gonads have enlarged and begun secreting certain steroid compounds. Second, the removal of the gland was attempted, with the expectation that the gonadectomized individual would cease singing and mating. The third step, the reintroduction of an extract and reestablishment of the behavior, led to the crucial *quod est demonstratum*.

Simple as these three steps may seem, the answers they provide are far from complete. Consider the example of the songbird. It is not entirely true that the injection of a gonadal extract or implantation of the gland will always produce an effect. In canaries, it is true, injection of the male gonadal hormone, testosterone, will induce singing in both males and females. But in Australian zebra finches, only males, not females, will respond by singing (Nottebohm, 1970). Clearly, hormones alone cannot control behavior; they interact with other structures, and only in certain cases produce an effect. In the canary, four areas of the brain are known to be involved in vocal behavior, and these are larger in males than in females. In zebra finches, only males have four areas; the females have but three.

Suppose, however, testosterone is injected into embryonic zebra finches. Might the presence of male hormone then induce the development of the otherwise absent fourth area? This experiment has not been performed, though we have a great many examples of hormones that occur or are administered prenatally influencing the development of particular organs. Indeed, the feminization seen in some mammals due to the administration of estrogenic (female hormone) compounds, has caused a considerable stir among clinicians. Many meat animals are given estrogens in order to make them gain weight more quickly. If the residues are consumed by pregnant women, their fetuses may be subject to unusually high estrogen doses. In short, even given the enormous variability in responses from one species to the next (rats, rabbits, and guinea pigs all differ), it can still be stated that hormones influence embryonic development. The resulting structures, including particular regions of the brain, may then show changed sensitivity to the presence of the same hormones in adulthood.

Feedback occurs in other contexts, too. We have seen that the simple notion of a hormone affecting a target organ had to be modified by recognizing that organ development is itself responsive to hormones, and, in turn, determines later responsiveness. External factors also influence a hormone's influences, for example, the social environment. When a male ring dove sees its mate, its prolactin levels rise (prolactin is involved in brooding and feeding behavior). Day length is another external factor known to influence hormonal output, and, in consequence, behavior. Migratory behavior results from changes in day length, which stimulate cells of the pituitary gland, whose secretions influence the gonads, which, in turn, produce hormones affecting areas of the brain. Interestingly, such light-induced changes also influence habitat choices. Birds under short-day conditions choose different habitats than birds under long-day lighting. Humans, too, often undergo depression under short-day, low-light conditions. This, and internal pacemakers are presumed to result from light-induced changes in hormone activity (Czeisler *et al.*, 1989).

2. HOW DO HORMONES ACT AND INTERACT?

Understanding the mechanism of hormonal action, as noted, is made more complicated because both physical and social factors may be influential. The behavioral changes one often sees when mammalian populations grow are due in part to the mutual stimulation of adrenal function. A few rats in a room will have relatively small adrenals. Crowd in more, the adrenals grow, and the animals' behavior changes.

The doyen of the study of hormones and behavior, Frank Beach (1948), suggested four mechanisms whereby, theoretically, hormones could modify behavior by:

1. changing the overall level of the organism's activity
2. altering the structures used in making particular responses
3. changing the threshold (sensitivity) of sense organs
4. directly stimulating particular neural centers

We now have examples of all of these, and furthermore know a bit about the many ways the mechanisms interact.

Consider the domestic goat, an animal we have used extensively in our own research. (Besides being suited to the purposes of our studies, goats make delightful pets, are more easily housetrained than dogs, and provide milk and cheese). A pregnant doe, shortly before parturition, becomes restless and separates herself from the herd. It is a fair assumption that this behavior results from the hormonal changes that trigger birth, though in this case, we have no firm evidence. We do have firm evidence, however, that as a result of the cervical stretching the emergence of the kid produces, there is a reflexogenous release of the hormone, oxytocin, from the midbrain. The presence of this hormone transforms a doe uninterested in or hostile toward a kid into a loving mother. The same effect can be achieved even in virgin does merely by manually dilating the cervix in a manner that mimics normal birth. Transfusing the blood of a doe in labor into a virgin will achieve the same result. Injection of the hormone itself seems to be effective only if the site of the injection is the third ventricle of the brain. In rodents, even males act like loving mothers when so injected (Pederson and Prange, 1979).

Once the doe has licked and nuzzled her kid, even if only for 5 minutes, a strong and lasting bond is forged between them. However, if this initial contact is delayed more than 30 or so minutes after the oxytocin's initial release, the doe will be uninterested if not hostile to her kid. The kid's presence when she is under the influence of the hormone alters her behavior so that she remains loving even after the oxytocin has disappeared from the bloodstream (Klopfer, 1971).

There appear to be two different mechanisms by which the oxytocin acts, and both may be involved. One likely possibility is that the hormones change the sensitivity of the olfactory receptors, allowing the doe to perceive alluring scents of which she is normally unaware. Once these have attracted her to her kid, other cues bind her to it. We know from our own experience, after all, how potent short-lived olfactory experiences may be. We also know that olfaction is remarkably sensitive to hormonal states, which is why young women are often unreliable wine-tasters; their sensitivities vary from one end of their menstrual cycle to the other. Responsiveness to newborns could very well be enhanced by similar hormone-induced threshhold changes.

The alternative possibility is that the hormone triggers activity in particular neurons, which in turn activate the motor patterns in question. The fact that oxytocin, but not other substances, specifically induces maternal behavior in male rats when injected into the third ventricle of the brain, but not when injected elsewhere, suggests this mode of action. The two possibilities are obviously not mutually exclusive. In short, the role of hormones in the control of behavior is not merely important, but complex.

Another and probably the most famous example of complex interactions was provided by a pioneer in this field, D. S. Lehrman (1965). He showed that in doves the production of hormones was induced by social stimulation, for instance, by the sight of a potential mate. Lehrman went on to postulate (and later demonstrate) this sequence.

After mating and egg-laying (all influenced by both social stimuli and hormones), another hormone alters the skin so a bare brood patch develops on the belly. The sight of eggs in a nest hastens its development; remove the eggs, and no brood patch may appear. This patch is apparently irritable, and contact with the smooth surface of the eggs appears to be pleasurable, inducing the bird to incubate. This incubation posture, in turn, stimulates the mate to be more ready to incubate. The feedbacks continue.

A final word about the relation between hormones and behavior may make these complexities seem more reasonable, if not simpler. After all, it does seem peculiar that not only do hormones cause behavior, but behavior causes the secretion of hormones. That final word has to do with the development of hormones during early embryogenesis. The glands that secrete hormones come from cells that can be traced back in development to one of the three primordial cell layers of which the very young vertebrate embryo is composed. This is also the cell layer from which nerve tissues, including the brain, develop. Neurons and endocrine glands are thus closely related, so it is not surprising that they continue to interact. Indeed, nerve cells themselves secrete hormone-like substances (neurohumors), which serve to transmit nerve impulses. Adjoining nerves do not make physical contact with one another. They are separated by a minute cleft. This is bridged and a nerve impulse is propagated by these neurohumors. Thus, functionally and developmentally, nerve cells and endocrine cells have behavioral similarities, so interactions might well be expected.

D. The Inner Life of Animals

1. UMWELT

Early this century a German biologist, Jacob von Uexküll, (1909) introduced the term Umwelt to his colleagues. Nominally, Umwelt means the environment, but what von Uexküll intended in his use of the term was "the environment as perceived by a particular animal." If you ride an equestrian cross-country course, you know that red and white flags mark the course, and one color must always be to your right and one to your left. Smaller blue or orange flags in front of or behind each jump designate the penalty zone where a fall costs points; you would see those colors. How different it must appear to horses, however, for these animals lack color vision. The different flags to them differ only in shades of gray. Their visual Umwelt differs from ours in this respect, though the environment itself is the same for both horse and rider. In the case of bees, which do possess color vision, the scene looks still different. Bees' vision sensitivity extends to the ultraviolet. Thus, two white flags or flowers that are indistinguishable to us or to our horse could appear very different to a bee if they contain dyes or fibers that reflect or absorb the sun's ultraviolet rays differentially.

How do we know that animals see the world differently? How do we know what it is they see, or smell, or feel, or taste? Can we get under the skin of different kinds of animals and perceive the world as they do?

von Uexküll's genius lay not merely in the postulation that different kinds of animals have different perceptual worlds, even when they live side by side, but in recognizing that clues to the character of that world are provided by their sense organs. Unfortunately for von Uexküll's reputation and influence within the discipline of animal behavior, his views were linked to a strong vitalist (nonmechanistic) outlook. Students of sensory systems tend to be mechanistic in their outlook, and so were uninterested in the philosophic overtones in von Uexküll's writings. Why not just study sensory structure and function and its relation to behavior, they argued, without making (unnecessary) inferences as to mental states and desires? Their successes with this approach are unmistakable. Two are described below, the analysis of the perceptual worlds of bees and bats.

2. BEES AND THEIR WORLD

Shortly after the First World War, the German biologist Karl von Frisch commenced what was to become a lifelong study of the sensory capabilities and perceptions (including the language) of honey bees. Many years later the importance of this work was to be recognized by the award of a Nobel Prize (shared with Niko Tinbergen and Konrad Lorenz).

Bees easily learn to take sugar water from a dish, and, as earlier workers had learned, can be trained to appear at a particular time of day. This is a great convenience for the experimenter. von Frisch used these traits to investigate their capacity for color discriminations. If the bees were trained to feed only from a dish set upon a blue background, they could later be offered dishes set on an array of grey backgrounds of different brightness. The bees were never confused, however; they clearly saw blue. In like manner, von Frisch mapped the full color sensitivity of bees and discovered that they see more colors than we do.

His next observation grew from the color experiments. His trained bees recruited additional foragers from their hive, and these bees arrived so swiftly after the feed bees had returned home that they must have made a beeline from hive to food source. "How did they know where to go?"

By watching returning scouts in a glass-covered observation hive, von Frisch noticed that the scouts danced in a circle when the food source was close at hand, but in a figure 8 when it was more distant (over 50 meters). This figure was really two circles connected by a straight line. Hours of observations finally led to the realization that the straight line made an angle with the vertical identical to the horizontal angle made between sun, hive, and food source. Thus, if the straight line part of the dance was vertical, it indicated that the food source lay directly towards the sun; if at right angles to the right of the vertical, then 90° to the right of the sun. The distance of the food source was related to dance frequency: the more

distant the food, the slower the repetition rate. The liveliness
of the dance indicated the abundance of the food or its rich-
ness, while the scents carried by the dancer identified the nature of the food.

There may be additional information provided by the
sounds emitted during the dance. Wenner (1967) believes
from his studies that the vibrations produced during the
dance are perceived through feet and antennae and provide
clues to distance.

Bees still find their way to a food source when the sun is
occluded by clouds, provided a patch of blue sky is visible.
This led von Frisch to the suspicion that not the sun itself but
some pattern it produces is the orienting cue. When he juxta-
posed a polarizing filter between sky and hive, he found he
could make the bees rotate their orientation as he rotated the
filter; the bee sees polarized light. The physics of this phenomenon is now well understood, as are the neural mechanisms involved. von Frisch's discovery established the existence of visual dimensions of our world to which humans are
blind. Even crabs and shrimp see planes of polarization,
which only fancy optical analyzers reveal to us. Furthermore,
the polarized patterns in the sky change as the sun steadily alters its position during the course of a day. Obviously the
bee's image of the sky is not static but is time-compensated.
A food source that lies at one angle at noon will be elsewhere
at 2 PM. The bee's perceptual world is evidently dynamically
structured.

3. AND THEN THERE ARE BATS

The remarkable ability of bats to avoid obstacles even in
the dark was originally demonstrated by an Italian investigator a couple of centuries ago. He strung fine wires from ceil-
ing to floor of a large room, just far enough apart that a flying
bat could go between them only by either briefly folding its
wings or turning sideways; the outstretched wings would
touch the wire and cause a bell, attached to each wire, to jin-
gle. Thus, even in absolute darkness it was possible to determine that flying bats could avoid these thin obstacles.

Spallanzani, the investigator, had demonstrated that bats see without light (Griffin, 1958).

The next step was not taken until the winter of 1938 when a Harvard undergraduate asked the famous physicist G. W. Pierce to lend him equipment that detected very high sound frequencies (ultrasound) beyond the range of the human ear (which is generally limited to a range of 20 to 20,000 cycles per second). The student, Donald Griffin, together with his friend Robert Golambos, thus solved a centuries-old mystery. Bats avoid obstacles by emitting pulses of ultrasound whose echoes define objects around them. If the bats' muzzles are tied shut, they cannot emit these pulses and will not fly. If thrown into the air, they collide with obstacles when forced to fly. If an ear is stopped with cotton, the same thing occurs.

Griffin has parked his station wagon near bat caves, an oscilloscope (and generator) on the tailgate. At the end of a long umbilicus from the oscilloscope stands Griffin, holding a microphone with parabolic reflector, listening in to the chatter of the passing bats. As he throws a pebble in front of a nearby animal, it veers towards it, and the oscilloscope screen shows it is emitting a stream of sound pulses. Then, it veers away. When a moth crosses its path, the pulses also pick up, but this time the bat pursues and plucks its prey. Evidently the echoes from soft and hard surfaces can be distinguished by the bat.

While Griffin considers himself an experimental naturalist, much of his work has been performed under the carefully controlled conditions of his laboratory. There, he and his students have been able to precisely characterize the pulses of different species of bats, to discover how they produce the sounds, how they receive them, what information the echoes can provide, how they distinguish the echoes of their sounds from the pulses of bats flying directly beside them, and myriad other details. Whales and whirligig beetles, birds and some mammals, they have learned, also employ ultrasound sonar. Nor has the applicability of the principles involved been lost on them for the development of prosthetic eyes for blind people. The discovery of a world of ultrasonic echoes is among the most dramatic demonstrations of von Uexküll's dictum that the perceptual worlds of animals may differ drastically from our own.

4. OTHER SENSES

The studies of von Frisch and Griffin, while seminal, no longer stand alone. A host of unusual (to us) sensory capacities and perceptual worlds have been uncovered in the last several decades. Some of these capacities are but amplifications of senses common to us. Owls can use their eyes in light too dim for us, and hear sounds more faint than those we usually detect. This is a necessary talent if they must feed upon night-active mice. Others use sense organs we share, but in a different fashion. Bees, for example, perceive the difference between kinds of flowers not by their shape, but by the visual flicker induced by the edges of the bloom. Finally, there are senses we do not share, for instance the ability of some fish to produce an electric field around themselves and then to detect (and discriminate) objects on the basis of the kinds of distortions they induce in that electric field. The ability of pigeons to detect geomagnetic fields, discussed earlier, is another example of what we might call a cryptic capability. von Uexküll's concept of the Umwelt (the world as the animal perceives it) is finally receiving the attention it deserves.

E. Social Behavior

Certain of Darwin's arguments concerning organic evolution rested upon the assumption that there is competition between organisms for a limited supply of resources such as food or nesting sites. Competition, of course, is almost never expressed directly through combat, but proceeds indirectly. Its outcome is seen as differences in proportionate contributions to the gene pool of future generations. A more anthropomorphic view of competition, however, proved generally pleasing during the mid-nineteenth century. Thus was born the un-Darwinian concept of "nature red in tooth and claw," a concept that implied direct aggressive competition and was scarcely what Darwin described. As Darwin's theories of natural selection and evolution gained ascendancy, Darwinism was increasingly interpreted as providing a natural sanction for current social and economic doctrines. "Survival of the fittest" became an excuse for the maintenance of ruthless social practices. This in turn led to an increasing distortion of the picture of the animal kingdom. Interspecific and intraspecific relations were seen with emphasis on all their ugliest details. The romantic notions of an earlier breed of naturalists and philosophers (e.g., Goethe, Rousseau) were rejected in *toto*. Far from being a peaceful kingdom, animate nature was brutal, selfish, and bloodthirsty. For man to be less, to give expression to altruistic impulses, was for him to deny his true nature; this was apparently a common nineteenth century view.

Businessmen and economists, of course, no less than politicians, rarely took the time to observe real animals as distinct from those they imagined prowling the pages of Darwin's *Origin of the Species*. Those naturalists who were more concerned with accurate observations than with justification of a particular economic order soon attempted to dispel the "realists'" illusions. Espinas, in 1878, published his *Des Societes Animales*. In this work he stressed the near-universal occurrence among animals of social organization and some form of communal life. Far from unrestricted and direct competition's being the ruling principle, Espinas found countless examples of social or altruistic behavior. His work should have made an impact, but it apparently failed to do so. The extreme individualism of the late nineteenth century made notions of cooperative organization repugnant. Just as "nature red in tooth and claw" justified the prevalent mores, a doctrine stressing the universality of social organization appeared to remove that justification. In any event, Espinas failed to convince most of his contemporaries, and it is largely in more recent works that we find him cited.

At the turn of the century and in the decade following, a series of studies of social behavior appeared. Some of these were concerned merely with a taxonomy of social forms, as in Deegener's (1918) classification of different levels of social organization. Others sought to prove the existence of a universal social drive or cooperative instinct. The most famous of these last is surely Prince Kropotkin's *Mutual Aid*, published in 1914. As a collection of heartwarming animal anecdotes, it is superb. As a serious contribution to animal sociology, it is somewhat lacking. Kropotkin was as determined to find a biological justification for a benign, cooperative social order among humans as were the economists of an earlier decade to find support for policies of *laissez faire*. Thus, he rather uncritically assembles those tales favorable to his thesis.

In the decades that followed, studies of social behavior developed a more objective character and were less often reflective of a particular political bias. Unfortunately, this did not mean that polemicists ceased to use particular studies to justify their own beliefs. While the studies of sociality, especially in insects, by scientists such as Allee, Emerson, and Wheeler stand out as landmarks, the questions they addressed continue to confront us today. What determines group structure? Why do groups form? How does social behavior evolve, and how does it develop?

Animal aggregations cannot all be termed societies. When a line of cars moving at high speed along a main road enters a town with reduced speed limits, the vehicles will clump. The clumping cannot be attributed to any social drive of either the cars or their drivers. Similarly, wherever local environmental or microhabitat conditions are such as to cause a reduction in speed of movement, an aggregation will appear. Such aggregations may have survival value, but they require no overt responses to one another on the part of the participating organisms. Nothing is needed, in short, other than a difference in movement dependent on whatever factors act as the entrapping agent. In addition to such aggregations, clumping of animals may result from a directional response to a specific environmental condition. If wood lice move toward regions of maximal humidity, and if areas of high humidity are localized, clusters must form. The resulting clumps are termed tropistic aggregations. As with other aggregations, they need involve no social interactions, merely individual responses to the same environmental stimuli.

Aggregations may also result directly from the response of one animal to another. A male attracted to a female, or parents attracted to their young, or young attracted to one another, provide examples of aggregations dependent on animal-centered rather than environment-centered signals. This should not be taken to mean that these factors operate entirely independently. For instance, in the Gilboa Mountains of Israel, herds of gazelles, *Gazella gazella*, graze on slopes of the broad wadis. To the east, visibility extends for several kilometers. There are relatively few visual obstructions; grass is short, brush or trees are generally confined to a narrow strip in the wadi bottom, and rocky outcrops are few. A few kilometers farther west, the wadis are narrower, their sides more precipitous, impediments to distant vision much more frequent. Preliminary observations of the gazelles suggest that these western herds generally move in linear fashion when alarmed. The animals are widely spaced, and distinctive leadership is shown by the first and last animals. In contrast, the more easterly herds more often move along a broad front, with no particular animal obviously setting pace or direction. Although our data on these gazelles are as yet incomplete, they do suggest how profoundly sociality and substrate interact (P. H. Klopfer, unpublished observations).

First, it will be of interest to consider the variety of ways in which these social aggregations may be organized. Three categories of relations can be distinguished.

1. There are mateships, which involve a pair of animals whose young disperse, leaving their parents after attaining maturity. Mateships may show varying degrees of exclusiveness and stability. Among some species of bears, for instance, the relationship is monogamous and seasonal. It is also solitary in the sense that the mated pair is not part of a large aggregate. Alternatively, a permanent though solitary and monogamous relationship may develop, as is thought to be the case with the rhinoceros. Mateships may also be polygynous and seasonal, as with deer, or permanent.

2. A more complex organization exists when the social unit is the family, a group that may include several adults and the young of previous seasons. This is the case in a great many birds and mammals.

3. Finally, we come to herds or packs, which are often most elaborate in the complexity of the social relations they

exhibit. Herds may include smaller groupings, either families or mateships, hence the increase in the potential diversity of social relations. The red deer, studied in loving detail by F. Darling (1937), have long provided the classical example of an intricately organized herd society which, in turn, is subdivided into smaller groups consisting of a mature hind (adult female) and several followers, generally the offspring of the previous three years. Both in the late spring, at calving time, and again in the fall during the rut, major realignments occur. Extremes of weather, severe insect infestations, or other disturbances add to the factors that may temporarily shift the organization of what is in essence a highly stable social structure.

The organization of animal societies can be attributed to four principal kinds of behavior patterns. These consist of the behavior associated with territoriality and the maintenance of individual distance, with dominance relations, with leadership, and with parental care and mutual stimulation. With the possible exception of the rigidly demarcated insect societies, it is possible to arrange social organizations according to the degree to which one or the other of these patterns predominates. Insect societies are a distinct phenomenon because of the unusually close genetic relationships between members. Bees or ants are more akin to the cells of a multicellular animal.

Causal references to the active defense of a certain area by an animal are found in many early natural histories. Credit for the emergence of the concept of territoriality in its modern form, however, is generally accorded the English ornithologist, Eliot Howard (1920). He noted that early in the spring, the males of many species of songbirds restrict their singing activities to particular trees, shrubs, or fence posts. The areas enclosed by these outposts are rarely if ever traversed by other males of the same species. Thus, Howard concluded, conspecific males divide their breeding grounds into territories each of which is occupied by only a single male. Later on, females, presumably attracted by the advertising song produced by the males as they make their rounds, may join a male in the patrolling of his territory. It will be recognized that there may be innumerable variations in this pattern from one species to the next. These variations may include interspecific differences in the size of the defended territory, the degree to which intrusions by other birds of the same or different species are tolerated, and so on. Sometimes the defended area moves with the individual. For such situations the term individual distance has been used, although the basic phenomenon appears to be the same. In either case, a periphery is created at which contact with other individuals may occur, which provides opportunity for other kinds of relations than those involved in nest defense. At the edge of the site, the tolerated approach-distance of other individuals shrinks well below the limit obtaining about the nest, for example, and absolute dominance gives way to more fluid interindividual relations.

It should be noted that an animal's territory is not identical with its home range. This last term refers to all of the area in which an animal passes its time, while territory is reserved for that part of the range which is defended against intruders. Although the two may be coextensive, generally the home range is considerably larger, and the home ranges of many individuals may overlap (for more on territoriality and its evolution, see Klopfer, 1969).

Dominance is inferred whenever one individual is able to chastise another with impunity. The ability to be dominant is generally thought to be most frequently a function of sex, size, and physical condition, though by no means is it invariably so. Among the Alaskan fur seals, possession of a female harem may be decided in favor of the strongest, heaviest male (so long as he remains the strongest), but among other territorial beasts the original possessor of a territory may remain the alpha or superordinate animal, almost independent of his physical attributes.

In cases of extreme territorialism, where each individual or pair has exclusive possession of a tract of land, social behavior will be relatively limited. Absolute dominance by the possessor over any intruder is the rule.

In other cases, such as the gulls, rather small nesting territories are defended, thus forming dense aggregations on the available breeding grounds. In addition to their individual territories, gulls have communal loafing sites. The psychological advantage that accrues to the defender of his own home is not unique to *Homo sapiens.*

Dominance is not a constant condition. Among some higher primates (e.g., baboons), for instance, an estrous female assumes the rank of her male consort. Once he loses interest in her, she reverts to her original status, which, in primates, is generally beneath that of the adult males. Nor is dominance independent of the context in which it is displayed. An experimental study of dominance among bushbabies, *Galago crassicaudatus,* led to the conclusion that the particular index used to measure dominance determined which individual was dominant. The concept was too heterogeneous to be meaningful (Drews, 1973). Others flatly reject the concept as anything besides a shorthand description of certain behavior patterns, proposing instead that more utility lies in a concept of subordinance.

The notion of dominance as a major phenomenon of social organization is largely attributable to Schjelderup-Ebbe (1922). The highest ranking animal can dominate all those beneath him, the second highest, all but his superior, and so forth. The positions in the hierarchy may be interchanged, of course, as one animal sickens or another grows more vigorous. In other species, such as pigeons, the linear hierarchy is not absolute, giving way to a more changeable peck–dominance relationship, such that A usually dominates B, but not to the extent of being totally immune to retaliatory pecks.

Despite differences in the degree, stability, and nature of the dominance relationship, the establishment of a convention of precedence goes far to assure the transformation of a mere assemblage into an organized society.

Leadership represents an increasingly important phenomenon in mammalian societies whose structure persists over long periods, in contrast to the seasonal nature of most bird flocks. The leader is the individual who determines the direction of rate of movement of a group or who sets its moods, initiating alarm or feeding behavior. Leadership of a group may be divided so that different animals have different roles. Among the red deer, the males may act as leaders of their harems during the rut, but should an alarm be sounded, they will depart by themselves, leaving harem leadership to an older hind. Similarly among reindeer, the animals who act as leaders in giving notice of danger are drawn from a peripheral group, which does not necessarily include those animals that set the pace. In both of these cases leadership is fairly stable. Among primates this is apparently also the case, but because leadership is closely linked to dominance, a change in one leads to a change in the other. Finally, leadership may in some forms be unrelated to either dominance status, spatial position in the herd, or astuteness. In both schools of fish and groups of ducklings, it is apparently suddenness or directedness of movement that inspires leadership. Whichever animal heads off most determinedly draws the crowd along. The parallels with human situations are amusing, if not altogether significant.

Many of the earlier treatments of animal sociology focused upon the advantages to the organism of the abandonment of a solitary existence. If such advantages are real and outweigh the disadvantages inherent in social groupings, then one might expect persistent evolutionary trends in the direction of social behavior. Of course, not all organisms are social, or at least not to the same degree. It would appear, then, that just as some factors may favor the evolution of sociality, others oppose it. After considering the advantages of social behavior, we will therefore have to ask why all animals are not social.

Even the simplest groups, mere aggregations, may produce individual benefits. A group of animals provides itself with greater protection against the wind than can one individual. If toxins are present, the larger the aggregation the greater the dilution effect, and the less the likelihood that any organisms will die.

Further, some organisms produce wastes that condition the medium so that its capabilities for sustaining life are improved. Some may grow faster in conditioned water than in the unconditioned, pure substance. The conditioning of the medium may be purely physical as well. A larger group of prairie dogs (within certain limits) is likely to fare better than a smaller group, because it has more burrows available when a predator threatens. Burrow construction then represents an important form of conditioning which is clearly better performed by several organisms that by isolates. The fact that such advantages do accrue to even the least organized of aggregates is of importance when we consider the evolution of societies and social behavior.

In addition to providing a buffer against climatic extremes and promoting growth, aggregates may derive protection from predation. Protection may be a passive affair, as when predators are frightened by numbers of animals which they would eat if encountered individually. The protective effect of aggregation may be the result of an active response of the organisms as well. Though this generally implies a high degree of social organization, it need not invariably do so. Animals could secure protection *en masse* as a consequence of the more aggressive behavior of the peripherally placed males. Sea urchins may accomplish the same end by virtue of the physical barrier many spines pose. One of our colleagues described the behavior of a captive seal into whose tank a school of anchovies was released. On previous occasions, when small numbers of the fish were released, the seal rapidly caught and devoured them. The larger number of anchovies, which formed a dense school whose outlines approximated the shape of an animal as large as the seal, not only inhibited pursuit but also apparently frightened the seal out of its pool. Though anecdotal, this account does illustrate the power of the masses.

The protective values of the aggregation are also enhanced because aggregations afford the possibility of polymorphism. A division of labor with certain individuals specializing in defense presupposes the existence of an aggregate. The social insects, of course, provide us with the most striking examples of this phenomenon.

Passing from the protective values of aggregations, we may consider a second major category of advantages related to reproductive functions. First, there is the mutually stimulating effect individuals have upon one another, which may both synchronize reproductive behavior and facilitate its occurrence. In weaver birds, the social stimulus of nest building by some males (itself elicited by heavy rains) results in nest-building behavior by the males throughout the area. Similarly, after the arrival of the females, a certain level of social interaction during the nest building displays of the males is necessary for successful pair bonding. It has been proposed that, especially among colonial birds, a population size below a certain minimum will lead to inadequate stimulation and reproductive failure. Above this minimum, there is a decrease both in the latency for reproduction and in the period of the year over which the breeding season extends (the Darling effect). This last could be advantageous in reducing mortality from predators. The need for a minimal number of individuals and the progressive shortening of a colony's breeding season as colony size increases, have been vigorously disputed, but it is still a tenable hypothesis.

In some cases social grouping may increase feeding efficiency. Among wood pigeons, inexperienced birds seem to learn the identity and likely location of food from experienced birds during flock feeding. Social feeding behavior also permits discovery of the location of sporadically distributed food, which a single feeding bird might not discover. If the food supply is low, solitary individuals have a lower survival rate than do social feeders.

Social patterns can help control mortality by assuring density-dependent controls of total numbers. Consider, for example, that animal populations are capable of exponential growth. This growth leads to populations of such a size that depletion of major resources occurs. Thereupon, a catastrophic decline in numbers and, depending on local soil and climatic conditions, a permanent demise of the population may result. The cyclic changes in the abundance of the lynx in North America throughout the nineteenth century and possibly of snowshow hares and lemmings may be examples of such oscillations. In these cases, the mammalian prey or grasses upon which the animals depend do return, allowing recovery of the population.

In tropical lands, however, where denuded lateritic soils may be quickly leached of their scanty mineral reserves, a temporary denudation may not be followed by a recovery of either the flora or, as a consequence, the fauna. Under these circumstances, extinction or extirpation will follow. Thus, it is clearly of value for organisms to evolve mechanisms to limit population growth before such extreme densities are attained. When social aggregates exist, such mechanisms can come into play far more rapidly than when individuals live isolated and dispersed.

Some of these mechanisms are clearly dependent on social processes. The highly social prairie dogs engage in much contact behavior, mutual sniffing, kissing, and the like. As the number of young increases, their kittenish demands upon their elders lead to a considerable increase in contacts. Ultimately, their elders emigrate, much as a child-besieged father might seek the sanctuary of his secluded office. The peripheral areas to which migration occurs generally lack the well-consolidated system of clearings and tunnels. The emigrants are thus exposed to much heavier predation, which few of them survive. The protection of the habitat has been achieved, however. It is unlikely that these prairie dogs, because of a social organization permitting a control of population density, will ever show the extreme cycles (and thus be exposed to the danger of local extinction), as is the case with northern lemmings.

F. Evolution and Sociobiology

Most efforts to develop insights into the evolution of sociality have depended upon the social insects, ants, bees, and termites, in particular. They provide such a variety of examples, which run from one extreme to the other, as to allow maximal freedom in the construction of evolutionary progressions. This is of questionable use or relevance to understanding the origins of vertebrate social behavior. Some of the adaptive features of sociality have already been noted. Of course, there can also be disadvantages to being social. Studies of primates, and *H. sapiens* in particular, have recently led to fresh ideas on the dynamics of social evolution, which have been advanced as sociobiology.

Sociobiology deals in large part with human social forms or cultures, and attempts to explain their origins and development. Much of this is also applicable to nonhuman societies, though our human language competence obviously brings new forces into play. Other animals also have some form of language, and to that degree are then subject to the same selective forces. Still, the major thrust of sociobiologists has been to explain the most complex societies, those of our kind, so we will also deal with examples that pertain to the human animal.

According to one of its principal proponents, Edward O. Wilson, sociobiology is "the systematic study of the biological basis of all social behavior" (Wilson, 1975, p. 4). This bland definition should not produce controversy. After all, even the most exclusively human attributes can be said to have biological bases. Even a preference for a Lafite Rothschild over a Taylor's red wine depends on the ability to smell and taste, surely a biological basis. What has made sociobiology a red flag to some is the inadvertent transmutation of basis to cause. This was not initially intended by Wilson, whose original text was sufficiently cautious to satisfy most of his peers. A political controversy did emerge from the suggestion that, just possibly, some cultural differences might be related to genetic difference. The details of that controversy, punctuated by picket line choruses and physical assault, provide a fascinating picture of the interplay of empirical science and ideology. We can state briefly the implications of the notion that biological factors are crucial in the development of cultural features, and the criticism of this idea.

Two assumptions are primary in sociobiology: first, that there are heritable traits, including behavior. Behavior that is near universal, that is, present in most if not all cultures, is assumed to be heritable. Three oft-cited examples are ophidiophobia or a fear of snakes; incest avoidance; and the expression of emotions by particular facial expressions. These attributes, it is claimed, characterize not merely humans but also many other primates and some mammalian species. Incest-avoidance is said to occur even in quail. This purportedly strengthens the view that these traits have a genetic basis.

The second assumption is that the fitness of an individual has to be measured not in terms of its individual survival but in terms of the survival of its genes. This measure is known as *inclusive fitness*. Thus, since each of us has about 50% of our genes in common with each of our siblings, our fitness is the same whether we survive and reproduce, and two of our siblings die without reproducing, or whether they survive at our expense. The survival of two siblings at the cost of our life also does not affect our fitness, since the two sibs carry as many of our genes as do we. This argument is generally accepted by biologists, and is not at issue. But there are a great many problems with assumption number one, to which we will briefly allude.

First, the problem of identifying heritable traits is not an easy one to solve, nor has it been resolved in the case of any of the behavior patterns of interest to sociobiology. Differences in features of a bird's or cricket's song, or the courtship dance of a fly, can be related to genetic differences, it is true, but these are trivial cases with respect to the social behavior sociobiologists wish to explain.

The belief that an ophidiophobic propensity is a general primate trait correlated with the presence of poisonous snakes is based, it appears, on anecdotal evidence. A laboratory investigation failed to support the view that snakes per se are fear inducing in rhesus macaques (Joslin *et al.*, 1964). As for incest avoidance, that is a case of a term misleading us. In humans, sexual relations with individuals of a particular group may be forbidden, and are termed incestuous, but the proscribed group may or may not be more closely related than a group with which copulation is allowed. Curiously, sociobiologists ignore the fact that severe *social* sanctions against incest would hardly be required if there were natural aversions for mating with kin. Incest is a human cultural phenomenon, by definition, and thus should not be applied to other animals. The instances of so-called incest avoidance in rodents and carnivores can generally be understood in terms of the old adage, "familiarity breeds contempt," or at least lack of interest. Kagan and Beach (1953) showed that when young rodents develop a play relationship, this later interferes with reproductive behavior.

Finally, the ubiquity of certain facial expressions can be documented, but it has also been documented that there is no universality in the emotional label attached to them.

There is more to sociobiology than this, of course, and more in the way of criticism, too. The origin and development of social behavior—territoriality, communication, dominance, altruism, division of labor, leadership, etc.—are a fascinating and complex topic, for the study of which we do not yet have a single overarching theory or methodology, as illustrated by attempts to explain territorial behavior.

G. The Evolution of Territorial Behavior: An Example of the Problems of Extrapolation

Efforts to deduce the evolutionary history of behavior have often foundered on the fundamental misconception that behavior represents a discrete anatomic structure, or even the deterministic expression of structures. This may be partially caused by the constraints imposed by our language, which requires us to cast all of our statements in the form of subject and predicate. Thus, *it thunders*, though *thunder* is both subject and action in one. A behavioral example worth considering again is the act of pecking by chicks, which has been studied in several species by a number of ethologists. In the case of the young laughing gull chick, *Larus atricilla*, the pecking is directed to a spot on the tip of the parental bill. The pecking appears to be a simple and stereotyped response. It is, in fact, both variable and complex, being composed of a multiplicity of movements and choices, which can be variously recombined without the final outcome's being altered. The important characteristics of the parental bill in eliciting the response include figure–ground contrast, bill orientation, diameter, color, and rate of movement. The motor response of the chick involves a step forward, a sideways twist of the head, an opening of the bill, a lunge forward.

Analysis of the pecking could certainly be pushed further along the reductionist path. It is quite clear that in the forward moving of the chick's head, the same muscle bundles do not have to be involved each time. And even within one bundle different fibers doubtless fire at different times. Indeed, the more one approaches the molecular level in describing behavior, the more probabilistic and nondeterministic does that description become. So then we have to ask what is the behavior that is being genetically determined, that is capable of evolving? Must we postulate a separate gene for each motor element of pecking, suggesting that each can be independently inherited? And for the perceptual preference, is each attribute to be considered as having been separately coded on the genome? This is akin to the view that the second letter of every English word has a particular and unique function. The inheritability and evolution of behavior becomes more intelligible if we view the gene, not merely as a repository of data or a blueprint from which an organism can be constructed, that is, as an inchoate homunculus, but rather as an information-generating device, which exploits the predictable and ordered nature of its environment. This view has been championed most cogently by Waddington (1966) and can be summarized as follows. A segment of the helix (the chromosomal DNA) specifies a particular species of RNA which ultimately, in an appropriate environment and in concert with other proteins, leads to the synthesis of a particular enzyme, which in turn may repress or activate further synthetic activity by that portion of the helix, or repress or activate another segment. "Wheels within wheels," and all dependent as much on substances external to the helix as to the structure of the helix itself. Hormones, for instance, whose synthesis ultimately can be traced back to the action of particular segments of the alpha helix, are now known to activate genetic transcription at another portion of the helix. The transcription products may further feed back and regulate development. (Klopfer, 1969, pp. 59–60)

In short, behavior, including territorial behavior, is no more contained within an organism than music is contained within a radio. Both radio receiver and brain act as transducers with the further difference that the brain may be thought of as a feedback system that examines (and may adjust) its own output.

This implies that territoriality cannot be treated as a comparative anatomist would the homologous limbs of quadrupeds. (Indeed, it may be questioned whether the notion of homology is even particularly useful with respect to forelimbs.) Territorial behavior quite obviously serves a variety of proximate functions. There is no reason to assume that it reflects the expression of any particular neural structure. Its development is more likely than not to be considered similar in principle to that of the pecking response described above. The fact that wolves, robins, and men may all defend a plot of ground allows of no further inference.

The conclusion that territories are diverse with regard to not only size, seasonal stability, and function, but also evolutionary and developmental origins, is not one to encourage speculative attempts at generalization. Yet the diversity posed by territorial behavior has, in fact, been encompassed within a single theory. According to J. L. Brown (1964), territoriality is most accurately regarded as site-dependent aggressiveness. Aggressive behavior itself is likely to be favored in evolution when its employment increases the prospects for survival of offspring. The aggressive behavior can be linked with any of a number of necessities, such as defense of food, mates, mating places, nests, or other requisites for survival or reproductive success. Whether aggression actually is employed to defend any or all of these resources depends on the *dependability* of the needed resource. Dependability reflects the availability and accessibility of the resource to each individual, and the cost (in time and energy) of obtaining and defending it.

Too much aggression in the absence of a short supply of the disputed requisite would eventually be detrimental. Consequently, a balance must be achieved between the positive values of acquired food, mate, nesting area, protection of family, etc., and the negative values of loss of time, energy, and opportunities, and risk of injury. Where this balance may lie in any particular species is influenced by a great variety of factors. ...Within the population those individuals with the optimal balance of the genetic factors working for and against a particular form of aggressiveness...would [become] the norms for the population. (Brown, 1964, p. 167)

The character of the territoriality that evolves, if any does evolve, thus depends upon the economics of site-dependent aggression. Marine birds, which nest colonially, feed from the sea: For these birds defense of feeding areas is impractical if not impossible, but defense of a small area to be used for mating and nesting is indeed possible. Sage grouse, however, range widely in their search for food; for these birds the major cause of mortality appears to be predation during the juvenile stage. Thus there seems to be only a limited value in defending a feeding area; the major problem for the sage grouse is to remain inconspicuous while the chicks are still maturing, and this interest is better served without territorial defense.

Brown also calls attention to the vast but often slight differences in the character and extent of territorial behavior exhibited by closely related species. His work on the scrub jay and Mexican jay is a case in point. The former bird is aggressively territorial, and the latter, although it inhabits ecologically similar regions, maintains only weakly defended communal territories. A comparable situation exists for the red-winged and tricolored blackbirds of the West coast. The existence of differences of such magnitude in what must be recently differentiated species suggests that territorial behavior is a labile trait, very responsive to shifts in selective pressures and easily changed or even lost altogether (Klopfer, 1969).

A territory is an area defended against intruders of the same species and sex, though there also exist instances of defense against members of alien species, or communal defense of group territories. Further, territories serve a multitude of functions, from the provision of an exclusive food supply, to offering exclusive mating or nesting opportunities. It is apparent that territorial behavior represents a set of adaptions differing from species to species in both underlying causes and developmental mechanisms.

Is Man Territorial?

The variety of territorial types and their widespread distribution across the animal kingdom argue against the notion that this behavior has arisen but once and represents a homologous phenomenon among different species. There is no reason at all to believe that, for instance, the kind of private property ownership exhibited by some nationalities of *H. sapiens* is in any way an evolutionary derivation from the defense of the nest site shown by a few species of songbirds. Some writers would even assume that because a man defending his castle experiences an emotion he labels *aggressive,* the bird must possess similar sentiments. Territorial behavior in primates other than man is far less in evidence than it is among birds. As for the motivational and emotional concomitants of the defense of property by man, it is well to recall

Hinde's (1966) admonition: "It is one thing to say that a rat has moved into chamber x, quite another to say it has gone to the reward box."

The origins of territorial behavior are very likely as diverse as that behavior is itself. It is almost certainly a polyphyletic phenomenon, which is not to say that all of the many forms of territorial behavior may not have something in common. Brown (1964) has suggested that we consider territorial behavior as an example of site-dependent aggression, and then asks under what conditions is it economic for an animal to devote its time, energy, and resources, to aggressive encounters? The answer, as indicated above, depends largely on the density, dependability, and distribution of the resource the animal needs, relative to its time and energy budget.

> For example, a bird such as [the] Great Tit of Europe nests in hollow cavities, which are rather scarce but of which a pair needs only one. It is quite economic for Tits to spend a certain amount of time defending this limited resource. On the other hand, during the winter, it has been estimated that these animals must feed on one small insect every two and one-half seconds. The food supply is abundant but distributed in very small packages. The amount of the resource that is lost by diverting food-gathering energy and time to food defense makes this trade uneconomic. Food resources should not be defended and a feeding territory should not evolve in this species. Similarly, a bird that feeds on a fairly large parcel of food that has an even distribution might find it economic to defend a feeding territory, as, indeed, some owls do. But birds, such as gulls, whose food is also available in patches but with these dispersed so that there are areas of great abundance and areas of paucity would not find it economic to defend a feeding site. In short, the type of territory which evolves will depend on the economics of site-dependent aggression, rather than on the presence of innate territorial urges. (Klopfer, 1969, pp. 89–90)

What about early humans? Is there reason to believe that site-dependent aggression served their needs? Whatever the answer, it is unlikely to alter the fact that the presence of private property rights among some modern people is independent of past precedents. The human organism is too malleable to allow ancestral habit to provide an explanation for much of present-day behavior. There is probably little reason to believe early humans to have been territorial. The paleo-ecologic studies by Birdsell (1953) paint a picture of an organism dependent on the gathering of low-energy food over a large region. This is more comparable to the ungulates, mobile creatures that utilize low energy, abundantly available foods, than to wolves.

Unlike ungulates, however, early humans were probably not markedly seasonal in their breeding. Hence a seasonal rut and the accompanying territorial behavior would not have occurred. Thus, it is plausible that early humans were neither especially aggressive (in the sense of not tolerating the proximity of others) nor territorial. The defense of property that characterizes some people today would therefore not be regarded as an inevitable, biological heritage, as is the wont of some popular writers like the dramatist, R. Ardrey (1966). Their suppositions rest on both errors of fact concerning the distribution and variety of territories, as well as an assumption about underlying mechanisms that must be made explicit. This assumption stems from the concept of instinct developed by the great ethologist, Konrad Lorenz (1965). This concept can be described by a model consisting of a reservoir within which a fluid collects. As the fluid accumulates, the pressure builds until a valve is released, and the escaping fluid energizes a stereotyped motor pattern. The model corresponds to the mechanism underlying instinctive behavior, presumably including territorial behavior.

However, assumptions on the unitary nature of the mechanisms underlying territoriality are, at best, gratuitous. Even if territorial behavior did at some evolutionary stage become adaptive for man, anthropological studies (Mead, 1935; Benedict, 1934) cast considerable doubt on the constraints this would have imposed on later developments. Territorial behavior in humans is far more likely to prove a cultural phenomenon than the result of a fluid territorial urge accumulating within humans until it must, inevitably, gush forth.

H. The Problem of Animal Reason

With the growth of the Animal Rights Movement, the issue of whether animals can think and feel has achieved an urgency unknown in earlier times. This is not to declare that consideration for animals depends on their possessing consciousness, whatever that is. Arguments for preserving and protecting animals can be no less convincing if predicated upon purely human ethics. The academic furor over the minds and emotions of animals does have an immediacy lacking in Darwin's time.

There are several separate but intertwined questions implied by the rhetoric on animal rights: do animals think, do animals feel, are animals conscious, and are they purposeful? Even in the early nineteenth century, when systematic studies of animal behavior began to appear at an accelerating rate, two extreme positions were being defended. Mechanistically inclined workers saw animals as automatons, bereft of any inner life. Jacques Loeb characterizes this approach with his mechanical explanation of animal orientation. Others were prepared to endow animals with all of the human emotions and purposes, though to a reduced degree. Darwin, after all, argued for continuity in evolution. It thus was reasonable to assume that human mental traits were presaged in the mammalian predecessors to primates. A popular response of this difference was proposed by Lloyd Morgan. His famous canon enjoined his readers not to ascribe to an animal, faculties that are not essential for the performance of an act. That, of course, evaded the issue rather than resolving it. An act could result from purely mechanistic processes but still be accompanied, however unnecessarily, by consciousness.

So much has already been written on this subject by philosophers and scientists that we can afford to be brief. Is there any experimental evidence that bears on the question? Is the issue one that hinges largely upon definitions or is it experimentally demonstrable? Both can be answered affirmatively.

Cognition

Computers do not think; they perform according to preestablished rules. Sophisticated programs do allow them to alter the rules as the consequence of certain outcomes, and since randomization processes may be part of the program, the computer may act as if it is thinking. Assuredly, it is not. It lacks spontaneity, insightfulness, and inventiveness. All those attributes that characterize cognition are set preordained limits. To differing degrees, these limits appear not to be so fixed for some animals, particularly the primates, but also many other vertebrates. The evidence from studies of insight learning with chimpanzees or latent learning in rats supports what haphazard observations have long suggested: animals are able to see spontaneously the connection between a box, a stick, and a banana hanging out of reach. A rat that is simply allowed to explore a maze, with neither reward nor punishment, will learn something about its layout. The assumption of an internal representation of the external world and its perceptual reorganization here becomes the more economical assumption. This is not to say all animals can think or even that any always do so, even humans. It appears safe to bet that some animals think, at least some of the time.

Emotions

What about feelings? Dogs do not blush, but many dog owners insist their animals can be embarrassed. Could it be? A semantic issue here intrudes itself. What are emotions? Older physiological theories (for example, the James–Lange view) hold emotions to be particular bodily states. An unexpected apparition induces a reflex flight; the ensuing hormonal and muscular actions are perceived as fear. From this comes the adage, "We are afraid because we run." The credibility of this interpretation was greatly damaged by the demonstration that denervation of the sensory fibers from the organs which were the presumed source of the perceptions did not extinguish the feelings. Direct stimulation of the organs also did not produce feelings. The same motor responses were often associated with quite different feeling states, so the James–Lange view was soon replaced. Its chief competitor, the Cannon–Bard theory, proposed that emotions were the consequence of the activity of a particular part of the brain, the hypothalamus, which was normally inhibited by the cortex. Cats surgically deprived of their cortex would, on being provoked, act as if in an uncontrollable rage. Lesions in the hypothalamus, on the other hand, could prevent the appearance of such outbursts. Unfortunately for the theory, so could lesions elsewhere than in the hypothalamus. Thus, another area, albeit a diffuse one, the limbic system, associated with the forebrain and its projections, was selected as the *locus* of emotions. But here, too, not all the facts fit the theory.

The most promising definition of emotions does not localize emotions in a particular region of the brain. This notion, developed by Pribram and Melges (1969), suggests that an organism becomes aroused whenever perceptions that result from sensory inputs deviate from what the organism had expected. You took a spoonful of soup you thought would be hot and found it icy. You could then establish a new expectation with which your perception is integrated. You might decide this is a fruit soup, intended to be cold, or you could determine this to be an unacceptable input, and reject another cold mouthful. What we term emotion is the feeling that reflects the neural processes, which have been aroused and which seek to reestablish their *status quo ante*. The accompanying figure illustrate these subtleties. (See Fig. 1.)

This model allows a number of interesting predictions, which have been confirmed, so it continues to hold center stage in studies of emotions. If we provisionally accept it, then animals must have emotions, for they are clearly subject to neural arousal when stimuli change. This can be confirmed behaviorally as well as electrophysiologically, by measuring the electrical activity of the specific regions of the brain. What we cannot say, of course, is whether the animal's feeling state has the least similarity with any we experience. However, as we know from cross-cultural studies, there are many feeling states unique to particular cultures. Even identical situations can elicit qualitatively different emotions in different cultures. Our inability to qualitatively describe or appropriately name animal emotions should not surprise us. It is certainly acceptable to say our misbehaving dog is acting *as if* embarrassed. It would be entirely without foundation to say it was embarrassed, even though we admit to the probable existence of emotions in animals.

Consciousness

And what about consciousness? Again, a semantic issue intrudes, because no two philosophers have yet fully agreed

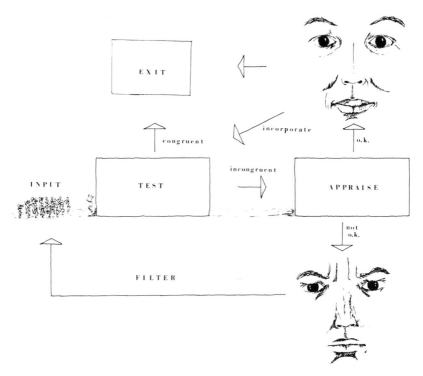

Figure 1   A theory of emotions. A diagram of the Melges–Pribam view of emotions. Incongruent or unexpected results are either incorporated into one's expectations or ignored. This process reflects what we term "emotion."

on a workable and inclusive definition. Definitions are arbitrary, of course; we can legitimately define a plane figure with four sides and four right angles as a circle, but this is not a useful definition because most people would name such a figure a square. Definitions must be useful. What do we need to know to make a useful definition of consciousness?

Two possible approaches come to mind. On the one hand we can define consciousness in terms of the most sophisticated human capabilities, those entailing foresight, purposiveness, self-awareness, language, and the like. Then, following Darwin, we can argue that such complex mental qualities are unlikely to have arisen in one quick step, but more likely through the acquisition and retention of many small changes. The continuity of mental evolution has been noted in learning ability, so why should it not be assumed for the appearance of consciousness? The issue of where in evolution consciousness first appears then becomes specious. It evolved gradually, continuously, and no line can be drawn separating the species that have it from those that do not.

A second approach is to define consciousness in terms of particular abilities that can be objectively discerned or measured, language ability, for example, or self-recognition. The latter is a convenient criterion, though, it could be faulted on the ground that consciousness ought to be more broadly defined. Operational definitions, while precise and useful, are generally narrow in comparison to nonoperational statements. We cannot have it both ways.

There is not much evidence for self-recognition. A study of a chimpanzee by Gallup (1970) illustrates one approach. While a tame chimp was asleep, Gallup placed a colored mark above its eye, being careful to avoid smelly or sticky cosmetics. When the animal was allowed to view itself in a mirror, as it was accustomed to doing, it promptly placed its hand on the mark. One cannot avoid the interpretation that this chimp knew what it should have seen, and reacted to the changed image.

Similar studies with other kinds of animals, other than primates, have not revealed similar abilities. However, it has been argued that recognition of their own breed, which can be demonstrated in dogs and horses, among others, represents a comparable phenomenon. Recently hatched chicks, too, preferentially assort with chicks of their own color phase if they have been reared together. In these studies, chicks that had been reared with an alien breed, however, did not develop a preference for being with the strangers. Thus, a degree of awareness of their own characteristics appears to be present in chickens and presumably other birds as well.

I. Animal Rights or Human Responsibilities

What does this say about the rights of animals? There is little disagreement, at least in theory if not in practice, with the proposition that cruel treatment of animals is to be condemned.

Whether imposed by a farmer, pet owner, or scientist, cruelty coarsens sensibilities, and is almost always counter-productive. Contented cows do produce more milk, and dogs that are not fearful are more readily trained. But another proposition has recently been raised that animals ought to be accorded legal protection against any kind of exploitation by man.

The issue of exploitation has many dimensions. First, consider the problem of exploiting, i.e., killing, wild individuals of species that compete with us for certain resources (lemurs that raid the orchards of impoverished Malagasay) or that we eat (whales). A strong case can be made to protect many of these animals when their exploitation threatens them with extinction, as is true for both lemurs and whales. (The lemurs are not so much threatened by direct killing as by destruction of their forest habitat.) The maintenance of ecological diversity is essential to human survival, important to human esthetics, and more arguably, a human moral responsibility so that future generations may prosper. Thus a conservation ethic is mandated even in the absence of a commitment to animal rights.

The second issue entails the exploitation of surplus stock (the excess deer, which resulted from the demise of large carnivores, in many eastern states of the United States) or captive animals reared for the purpose. Here, too, a case against such use can be made on ecological grounds without the necessity of invoking philosophical principles. In an ever more crowded planet, the inefficiency entailed in unnecessarily long food chains cannot be sustained. Plants capture but 1% or so of the incident energy from the sun; animals that eat plants utilize but 10% of their energy; those that eat those animals also gain but 10%, and with each link in the chain, another 90% is lost. Were humans to eschew beef for grain, only about one-tenth of the land area we presently use would be required for us to be adequately fed. Actually, the savings would be less dramatic, because few of us eat an exclusively meat diet.

Then we come to the case of the selective use of animals for particular purposes: dietary supplements not wasteful of resources (carp in farm ponds, for instance), or medical supplies (antibody production by horses), or scientific studies (including the training of surgeons or the development of medical procedures). Are these to be permitted under any conditions, some conditions, or not at all?

It is, we think, safe to say that the ecological arguments supporting conservation and vegetarianism do not cover these uses. Neither are there available suitable alternatives to the use of animals in all but a fraction of these cases. Dead animals will not do for the development of surgical techniques (how would you know if the surgery were successful?), even less for physical models. If a halt to these uses is to be made, it must be done solely on the basis of quasi-religious beliefs, and not from what we know about animal cognition and consciousness.

In earlier times, animals and their relations to people were viewed differently. In Europe during medieval times, it was not unusual for animals that transgressed the law to be formally charged and tried in court. In one celebrated case, rats that had destroyed a granary were first ordered to appear in court, then formally absolved of contempt for not appearing. French law excused a failure to answer a summons if to do so endangered the life of the defendant. The rats' court-appointed counsel successfully argued that the cats and dogs living between the court and granary did indeed pose a mortal threat to his clients (Evans, 1906; Klopfer and Polemics, 1988).

The view of animals implied by practices such as these is characterized by Lessley (1990) as "Animals are like us." This view has an alternative version, which allows for some actions (slaughter for food) that would be forbidden if practiced on humans.

At the Enlightenment, largely under the influence of Descartes, an opposing view gained prominence: animals are *not* like us; they are mindless (and soulless) automata. The weaker alternative to this view argues that animals should be treated compassionately because to do otherwise would contribute to moral insensitivity in interhuman relations.

Our hodgepodge of laws on animal rights reflects these seemingly irreconcilable positions. We do generally agree that the extreme positions, "Animals are just like us," and "Animals are totally unlike us," are untenable, but the gap between "Animals have the right to..." and "People have a responsibility to..." remains philosophically as great as ever.

Sometimes insights are gained by not directly confronting an issue. In this case, considering the views of cultures and religions other than those of the Western world may prove helpful. Neilhart (1959) portrays a contrasting view from the Ogala Sioux: "Everything an Indian does is in circles, that is because the power of the world works in circles, and everything tries to be round ... so long as the hoop was unbroken the people flourished."

This is directly relevant to the many environmental dilemmas that have resulted from human mismanagement and disruption of natural cycles, pollution, greenhouse effects, ozone holes, overgrazing, and deforestation. The ecologist (and one-time U.S. presidential candidate) B. Commoner has shown ecological problems cannot be solved simply by reducing issues to their component parts and addressing them separately (Commoner, 1971).

The multitude of interactions at and between different levels of the food chain demands a holistic approach. At the same time, the complexity of those interactions may surpass the analytical capabilities of the human brain. In some respects, poets with metaphor as their tool may be more successful at describing ecological processes than biologists with their instruments (Bateson, 1972). Thus a healthy respect for the natural world and a cautious approach to recipes for alterations or control seem called for. How better to promote such an attitude than by defending the integrity of animals. (Klopfer and Polemics, 1988, p. 107)

J. Suggested Reading

As was true for the earlier chapters, the most succinct summary (with many further references to work done up to 1975) and references to work cited by author only, is to be found in *An Introduction to Animal Behavior: Ethology's First Century* (P. H. Klopfer, Englewood Cliffs, New Jersey: Prentice-Hall, 1976). Two recent and readable summaries of the status of animal orientation and related topics are provided by K. Schmidt-Koenig in his *Migration and Homing in*

Animals (New York, New York: Springer, 1975) and *Avian Orientation and Navigation* (Cambridge, England: Cambridge, 1979).

The last word, or at least the latest word, on social behavior is to be found in *Social Evolution* (R. Trivers, Menlo Park, California: Cummings, 1985), in which a short theoretical treatment is given by R. Axelrod, a political scientist (*The Evolution of Cooperation*, New York, New York: Basic Books, 1985). And then, of course, there is the broad-ranging, rich and stimulating compendium by E. O. Wilson, *Sociobiology* (Boston, Massachusetts: Belknap, 1975).

Issues pertaining to animal rights are raised by D. R. Griffin (*The Question of Animal Awareness: Evolutionary Continuity of Mental Experience*, New York, New York: Rockefeller, 1976) and W. Paton (*Man and Mouse: Animals in Medical Research*, New York, New York: Oxford, 1984), Klopfer and Polemics (1988), and Lessley (1990).

4

Human–Animal Interactions

Now that we have a feeling for the history and concepts of ethology, we can begin to examine the behavior of specific animals. This section of the text focuses on domestic animal behavior, with a brief look at some of the commoner exotics and pets. It is aimed at the veterinarian, the student of animal science, the farmer, or the pet owner who wishes to gain some understanding of why and how his animals act as they do. We intentionally omitted a section on the behavior of laboratory animals, such as rodents, rabbits, and primates. This is not because of any prejudice against these creatures or the people interested in them, but merely because there are already several excellent texts dealing specifically with this group. Those who wish to learn more are urged to look at the list of suggested readings at the end of each section.

Why should we study the behavior of domestic animals? Aside from the purely theoretical interests that always appeal to scientists, there are a variety of more practical reasons. As vets, we try to help farmers realize their goals while promoting the welfare of the animals. As farmers, we are concerned with maximizing production while keeping costs within bounds. As pet owners and breeders, we strive for healthy, happy animals to share our lives. The achievement of all these ambitions depends on a good understanding of the needs of each species concerned, and not just physical needs. We ourselves know the tremendous impact our emotional state can have on our daily performance. The same can be true for our animals. For maximal production of milk, offspring, shiny coat, or rapid growth, a domestic animal needs more than just good feeding. Most producers know this, of course, if only through trial and error. Many of the practices of modern farming are based on informal experiments, and the farmer may not be aware of the reason for the success of a certain method. The point is, a basic understanding of the principles of animal behavior in general, and of certain groups of animals in particular, is of great importance to the management of animal production, health, and welfare.

The following section on behavior is divided into five categories:

1. Ruminant herds—cattle, sheep, and goats
2. Nonruminant herds—horses and swine
3. Carnivores—dogs and cats
4. Avian flocks—chickens, turkeys, ducks, and geese
5. Exotic pets—cage birds, rodents, reptiles, and fish

These categories, except for the last one, are based on the lifestyle of the animals in question. It may seem artificial to put horses and cattle in different groups; both are large, grazing herd animals, after all. And how are pigs similar to horses? If you look only at the standard U.S. hog and horse operations, you will see very few similarities. However, in considering behavior, we must consider evolution; we must go back through domestication and have a look at the animals from which our modern Hampshire hogs and Quarter Horses arose. What were they like? How did we get the animals we have now? In what ways are they different from (or similar to) their wild ancestors?

Most authorities make a distinction between *domestic* and *tame* animals. Taming is a process which reduces the fearful or aggressive behavior of animals toward humans. Domestication implies selective breeding and propagation of desirable traits (Wood-Gush, 1985). These traits might include high reproductive rates, certain coat colors, rapid growth, and so on. Psychological attributes, such as docility or learning ability, may be selected for as well.

Clearly, it takes many generations of deliberate selection to develop an animal that consistently demonstrates the desired traits. Recalling what we know about genetics, we might also expect to see the development of other characteristics not specifically sought, but which seem to be associated with the appearance of the selected traits. For example, a high incidence of hermaphroditism is seen along with polledness in certain breeds of goats. Polledness (the natural absence of horns) is a desirable condition from the point of view of dairy goat breeders. Hermaphroditism obviously is not, yet selective breeding to get polled goats has increased its incidence. Selection will not create new traits, but it does result in a new balance of both behavioral and physical attributes (Fuller 1969).

Surprisingly few species have been domesticated, worldwide. Why were some chosen instead of others? Some of the reason has to do with geographical location. People in the Middle East saw useful purpose in goats and sheep; South Americans turned to the llama and alpaca. In parts of the ancient world where there were no humans, the native species of animals were left untouched. But there is more than chance involved. Certain types of animals seem to have qualities that predispose them to successful selective breeding, and hence domestication (Ellis 1985). This list of qualities was first proposed by Hale in 1969, and has been only slightly altered by subsequent researchers. In summary, domesticatable species tend to live in stable, organized social groups. Young animals usually form firm bonds with their mothers (or other female caregivers), and are precocial in their physical development. These animals are herbivores or omnivores, and are very adaptable to environmental changes, desirable attributes in farm or companion animals.

The first animal to be domesticated was probably the dog, perhaps as early as 10,000–14,000 B.C. At this time, jackals, ancestral wild dogs, and wolves tended to follow nomadic tribes as they moved, scavenging scraps from their litter piles. It is not hard to imagine the transition from an annoying hanger-on to a useful addition to the tribe. Tame dogs can help drive off potentially dangerous packs of wild dogs, as modern farmers well know. The goat was the first agricultural animal to undergo domestication, followed rapidly by the sheep. Most sources place this occurrence at around 6000 to 9000 B.C. By 5000 to 6000 B.C., cattle, and swine in some parts of the world, were also an important feature of the herdsman's life. Horses were probably domesticated as early as 4000 B.C. Domestic fowl are more difficult to date; there are records of birds kept in captivity in China by 4000 B.C., but it is hard to say if these were merely trapped wild birds or if there was a true effort being made to breed them selectively. Similarly, tame geese appear in ancient Egyptian art from about 4000 B.C. The cat appears to be one of the most recently domesticated animals. It was certainly domesticated by 1500 B.C., and possibly even before that (Protsch and Berger, 1973; Zeuner, 1963).

In all of these cases, the source animal was a member of a wild species native to the areas inhabited by early humans, a species presumably possessed of those attributes we have previously discussed, preadapted to domestication. As humans turned from being primarily hunters and gatherers to an agricultural life, they began to cultivate animals as well as plants. Just as with the dog, it is easy to imagine the gradual development from captured wild prey to managed herds.

How much have our animals changed over the centuries? The phenotypic differences are clear; today's Holstein dairy cow bears little resemblance to the scrawny African and Asian cattle that are her ancestors. (See Fig. 1.) How much of her behavior has changed as well? This point is hotly debated by breeders and biologists alike. Physical and behavioral characteristics are intricately related in ways we do not yet understand; they cannot be chosen independently like cards from a pack. The consensus appears to be that domestication has altered the behavior of most species in certain respects, but not as much as we might expect. Exactly

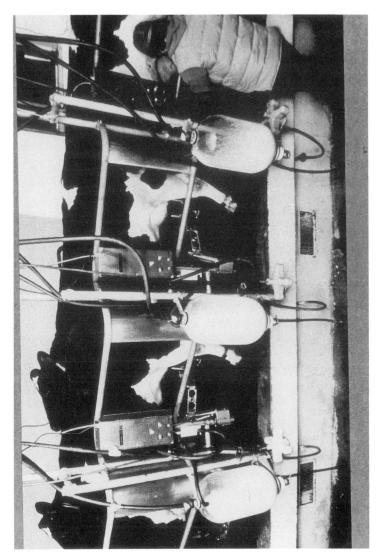

Figure 1 The modern Holstein dairy cow is a result of generations of careful selective breeding for milk production. (Photo courtesy of J. Fetrow)

what has been changed depends in part on what is being selected. For example, the modern laying hen is notoriously nonbroody in comparison to wild fowl (Craig, 1981). Humans have bred this bird for thousands of generations, selecting for maximal egg production over a maximal time. Obviously, this is not compatible with brooding chicks. Yet in other aspects of social behavior, the laying hen is remarkably similar to her wild ancestors. An understanding of the behavioral ecology of these ancestors and related wild species can be very useful in managing all of our domestic animals.

=== 5 ===

Domestic Animal Behavior

Most of our domestic species have been extensively observed at some point by researchers interested in behavior. We can look in a book and find out how many times per hour the average cow chews her cud, or how well young pigs can be trained to do various tasks (Hafez, 1975; Craig, 1981; Waring, 1983). This information may be useful, at least to someone who wants to know all there is to know about a particular species. On the other hand, it is often difficult to sift through the data to find those points most relevant to the farm (or other) situation. We have tried to select, for each group of animals, one or two major issues in behavior that are common sources of management problems. Each section begins with a short look at some wild relatives of the animals in question, with emphasis on the relation of behavior and lifestyle, and then proceeds to the domestic species. We do not mean to imply that the topics discussed are the only ones that should be considered when dealing with each species, of course. We cannot separate "behaviors" into neat little sections like oranges. It is not our intention to provide an exhaustive catalog of everything known about the behavior of each species. Our goal is to provide an understanding of the link between environment and behavior, and to encourage readers to consider this link and its significance whenever they are involved in the manipulation of an animal's environment.

A. Ruminant Herds: Cattle, Sheep, and Goats

Domestic cattle, *Bos taurus* and *B. indicus*; sheep, *Ovis aries*; and goats, *Capra hircus* all belong to the family Bovidae. This family includes their wild relatives, the African buffalo *Syncerus caffer*; Bighorn sheep, *O. canadensis*; and European and Syrian ibex, *Capra ibex* and *C. nubiana*. Deer, antelope, and musk oxen are also bovids. In some of these cases, the family resemblance is clear. But who would immediately deduce that the tiny duiker (an African antelope) is cousin to the lumbering moose, or the stately giraffe? Clearly, the process of evolution has wrought some major changes.

The first bovids appeared in the middle of the Tertiary period, about 70 million years ago. They were probably small forest dwellers, and evolved by adaptive radiation to fill unoccupied niches in the developing savannahs and grasslands (Estes, 1974). Some remained small and quick, nibbling on the short fine grasses. Others developed powerful jumping muscles to escape predators. It takes only a little imagination to predict the physical characteristics of a grazing animal occupying a particular type of habitat.

Behavioral characteristics can also be predicted to some degree. For example, forest-dwelling browsers tend to live in smaller groups than open-land grazers; a physical limitation is imposed by all those trees. Forest animals are more often territorial. That is to say, each individual, or occasionally each pair, defends a certain area against intrusion by other individuals. The size of the territory usually depends on the density of resources in the area, including food, water, and (because it is most often a male who will hold a territory), females. A territory is not the same as a home range; the latter refers to the entire area in which an animal, or group, may be found (Brown, 1975). Home ranges are typically larger than territories, and may be occupied by either territorial or non-territorial species. In the former case, a small portion of the home range is usually defended as a territory. Nonterritorial species are usually found in wide-open spaces, where it is not practical or necessary to defend a fixed space. For reasons of defense, open-land animals group into herds or flocks, which move as a unit in search of grazing grounds. If the grass growth pattern is seasonal, the herd may migrate with it, a regular, predictable movement, similar to that seen in certain birds. If the availability of food is constant within a season, the herd often displays a nomadic pattern, unpredictable roaming from place to place, depending on where the grass grows greenest. Of course, herds lucky enough to find themselves in regions where the grazing is lush at all times have no need to move far. In these cases, each herd in the area tends to occupy a loosely defined home range.

In general, the grazing ruminants are a gregarious lot, living in large herds of mixed sexes for at least part of the year. They are migratory, nomadic, or moderately territorial, with diurnal peaks of activity, usually at dawn and dusk. Home ranges may be extremely large, if the landscape permits. Most species in this group are seasonal breeders, although some of the Southern animals breed year round. There is nearly always some type of dominance hierarchy present, though its structure depends on a variety of factors. Males tend to be dominant over females once they reach adulthood (Estes, 1974).

The above generalizations, of course, apply to our domestic ruminants as well. We can get a better idea about the natural behavior of cattle, sheep, and goats by looking more closely at related wild species. In the case of cattle, *Bos*, the most appropriate example is the Cape buffalo, *Syncerus*. This large grazer exhibits the typical bovine social system, also demonstrated in free-ranging herds of domestic cattle. Males and females mix together in one large herd, composed of animals of all ages and social rank. In addition, males often associate together in "bachelor clubs," composed of several different dominance hierarchies. These hierarchies among males tend to be linear; that is, male A is dominant over all other males, male B is dominant over all males except male A, and so on down the line. Occasionally there will be cases of equal status, or where B is dominant over C but not D, but the general pattern is toward simple linearity. Dominance is established and maintained by displays of threatening behavior and sparring matches, where the contestants seemingly test one another's strength without actually fighting. Objective criteria such as body weight and horn size may be important factors in determining status among males (Mloszewski, 1983).

Females also exhibit dominance hierarchies among themselves, but these differ somewhat from those of the males .They also tend to be linear or triadic (A is dominant over B, B is dominant over C, but C is dominant over A), but there are many exceptions; the system is more complex. Perhaps the main difference from the male hierarchies is the permanence of status among females. Most researchers feel that dominance among buffalo cows (and other species, for that matter) is based on mutual familiarity within a stable herd. These animals spend a great deal of time in close proximity to one another. Calves are born and raised in the herd, and easily get to know all of its members, or at least the female members, who do not leave the herd at any time. The daughters of high-ranking cows tend to assume high rank themselves upon reaching maturity. (See Fig. 1.) This is a common pattern in troops of social primates (Chevalier-Skolnikoff and Poirier, 1977). It is not surprising to see it among social ungulates as well.

Sex differences in dominance hierarchies provide a good example of the relation of lifestyle to behavior, an issue which will arise time and again in the study of any animal. It is important to consider this relationship whenever one is working with animals, whether as an ethologist, farmer, or veterinarian. This seems like an obvious statement, but it is amazing how many times it is ignored.

The Bighorn sheep, *Ovis canadensis*, will serve as a wild model for our domestic sheep, *O. aries*. Most of what has been said about grazing animals applies to sheep as well as cattle. There are some differences in bovine and ovine social behavior, however. Males and females maintain separate flocks for most of the year, coming together briefly at watering holes and during the mating season. Thus the most com-

Figure 1 A cow's social status affects the relationship between her calf and the rest of the herd. (Photo by G. Honoré)

mon pattern is a relatively large, stable herd of ewes and their offspring, and several smaller bands of rams. The ewe herd tends to be based on maternal lineage; in other words, it consists of an old ewe, her daughters, their daughters, and so on. Once ram lambs reach subadulthood, they leave the maternal herd and join other young rams in a satellite band. As in buffalo, separate hierarchies exist for males and females, with adult males nearly always dominant over females. During the mating season, when the two groups come together, the most dominant rams do most of the breeding, a common pattern among ungulates. Immediately before this season, there is intense sparring and display activity among the males, as rank is established or reestablished. Most people are familiar with the spectacular photos of Bighorn rams engaged in these matches, which usually stop before any damage is done. However, serious injury can result if neither backs down and the confrontation proceeds beyond sparring to true all-out fighting (Geist, 1971).

Sheep, both wild and domestic, tend to form physically tighter and closer herds than do cattle or goats. As every shepherd knows, one animal can be responsible for the movement of the entire group. (See Fig. 2.) In Bighorn sheep, this leader is usually the oldest ewe, the one with the most offspring. She is not necessarily the most dominant ewe in the hierarchy (though she often is), and even rams will follow her lead when they are in the flock. This pattern has been observed in domestic sheep as well (Monson and Sumner, 1980).

The ibex, *Capra*, the same as our domestic goat, is probably very similar to the goat first domesticated several thousand years ago. Goats are truly browsers. They will graze, but they have broad tastes, and happily eat bark, leaves, and twigs that sheep ignore. This flexibility in diet is reflected in increased exploratory behavior, and, among ibex, reduced social contacts between herd members. Once again, behavior reflects lifestyle. In other respects, ibex herds resemble Bighorn sheep flocks, with maternal groups of females and kids, and separate bands of bucks. Male ibex are often solitary, especially at the start of the mating season.

We now have a feeling for the basic ecology and behavior of ruminant herds. How can this knowledge be applied to domestic ruminants? What are the implications for herd management and welfare?

In the wild, the size of the herd's (or the individual's) home range is largely self-determined. It may be affected by the number of other herds (or individuals) present, of course, and by physical constraints such as mountains and deserts. There is usually sufficient room to spread out enough so that no one feels crowded. Exceptions occur during times of drought, or fire, or other disasters that force animals into closer than usual proximity. Closer proximity means an increase in interactions between animals, which frequently leads to an increase in aggressive behavior. We are familiar with the concept of personal space in our own lives; each person is surrounded by this imaginary aura, which varies slightly in size according to conditions at the moment. We become distinctly uncomfortable when someone else invades our space, by standing too closely, or leaning toward us while speaking. Our first impulse is to back away from the intruder, increasing the distance between us to comfortable levels. If

Figure 2 Sheep, in small or large flocks, tend to bunch together and move as a group. (Photo courtesy of C. Uhlinger)

that fails, or if for some reason we cannot move, we may become agitated, perhaps even enraged. The most extreme response is to threaten the intruder (verbally or physically), and maybe even to give him a shove. This rarely happens, of course. We are constrained by the rules and customs of our society, which frowns upon such behavior. Animals also have personal spaces, but they are referred to as *individual distances* (Klopfer, 1974). The concept is quite similar to that just outlined for humans. Researchers have conclusively demonstrated that crowding, the reduction of individual distances, results in physiological signs of stress (Mason, 1968). The levels of corticosteroids and adrenalin both rise sharply in crowded animals. This physiological state tends to make individuals, especially dominant ones, more likely to attack. Under ordinary circumstances, the attacked animal, usually a subordinate, will withdraw from the attacker, increasing the distance between them. Under conditions of crowding, retreat is often not possible. Thus the attacked animal either fights back, despite its lower status, or suffers injury, or both.

The issue of crowding and its associated stresses becomes significant with domestic animals. Humans are in charge of the space available to range cattle, confinement-raised hogs, or dogs in a kennel. Many researchers have worked out the "minimum space required" for each animal (how much cow per cubic yard). There is a great deal of variation in these numbers, and in the criteria used to decide what constitutes a requirement. In grazing animals, of course, there is the pasture to be considered. An acre of grass will support only a given number of animals, no matter what you do to it. Given the gregarious nature of cattle, sheep, and goats, a herd placed on pasture of adequate size to support its numbers will probably have ample room. Problems with aggression and other stress-related disorders show up in confinement operations, or grossly overcrowded pastures. Stress and fighting can have serious consequences to the producer, whether his goal is milk, meat, or offspring. It may seem economical to put as many animals as possible into one barn, but in the long run, crowding is rarely cost-effective. Fertility rates and milk production in dairy cattle drop markedly when they are overcrowded. Normal social behavior is disrupted, with disastrous effects on mothering and dominance hierarchies in sheep and goats. The subordinate animals will get less to eat, either from limited access to food, or from limited time to feed in peace (Huntingford, 1984). One solution to the problem of increased aggression is to pack the animals in even tighter. In such cases, aggressive behavior declines, probably because the animals can barely move. There are obvious drawbacks to this solution, both from the point of view of the farmer and of his herd.

Solutions are difficult to find, because often the requirements of management go against the natural behavior of an animal. In a dairy operation, for example, calves are removed at birth and raised in individual hutches. Their mothers' milk is in demand, of course, and the lack of contact with other calves is supposed to reduce the spread of disease. This is all very logically thought out and planned, but how different from the normal lifestyle of the bovine species.

It is possible, however, for a farmer to both realize his production goals and improve the quality of life of his animals. In fact, the two go hand in hand. Overcrowding is one of the major problems on a farm, but it is usually not feasible for the farmer to provide more space or cut down the numbers in his herds. It is not even necessary to provide more space, as long as the space provided meets certain requirements. Thought should be given to the placement of food and water sources. There must be adequate room for subordinate animals to escape, if need be. Monotony, pure and simple boredom, also accounts for a large number of behavioral problems in herds. Some effort should be made to provide distractions, whether music in the milking parlor, or cinder-block mountains in the goat pen. Play behavior, either with others or solo, has been shown to be a significant part of the life of both juvenile and adult animals (Fagen, 1981). Opportunities to play ought to be made available on the farm. (See Fig. 3.) This can include such physical structures as the aforementioned mountains, or a butting board, or simply enough room to run and frolic.

Whenever possible, groups of animals should be allowed to remain together once they have established their own dominance hierarchies. Social stability is of great importance to

Figure 3   Opportunities to play, with members of the same or other species, are an important part of the social development of young animals. (Photo by P. Klopfer)

cattle, just as it is to children. Of course, bullies or trouble-makers need to be given special consideration; it may take some trial and error to come up with satisfactory groupings. As a general rule, the farmer and attending vet should try to recall what they know about the natural behavior of the animal in question, and then apply this knowledge along with a splash of common sense in designing management protocols.

A good example of this practice is given by the design of chutes to move cattle from place to place on a farm. Cattle are highly visual animals; they also have a natural tendency to avoid corners. They are not likely to willingly enter the traditional linear chute, with its straight lines and sharp corners. The resultant chasing, running, and confusion is very stressful to both the cattle and the herders. A simple solution is provided with the use of a curved chute. The cattle never see a dead end ahead, so they proceed readily through the entire chute, following one another closely (Gradin, 1978), and stress is thereby reduced.

B. Nonruminant Herds: Horses and Swine

Domestic horses belong to the family Equidae, along with zebras and wild asses. All equids are members of the Perissodactyla, or odd-toed ungulates. Domestic swine, wild pigs, peccaries, and warthogs all belong to the family Suidae, which joins the Bovidae in the Artiodactyla; one look at a pig's foot will assure you that it has the requisite even number of toes. Evolutionarily speaking, the Equidae and the Suidae are not very close, but they both differ from the ruminants in having a single stomach. One advantage of the ruminant four-chambered stomach is that the animal is able to take in a relatively large amount of food at one time and then digest it at leisure. Monogastrics like the horse and the pig do not have this option. Their stomachs are surprisingly small. In order to deal with the breakdown and digestion of grasses and other plant matter while still emptying the stomach at a rapid rate to make room for more, these animals have developed enlarged colons. Even with this increase in gut surface area, a horse must continue to take in grass at a fairly steady rate in order to meet its caloric requirements. Horses spend a much greater part of the day grazing than do cattle, for example. Pigs developed more omnivorous habits, which allow them to increase their caloric intake without increasing bulk; thus pigs need not spend more than about 25% of their time foraging.

Like the bovids, equids emerged from the forest many thousands of years ago and began living on the open grasslands in groups. This pattern is seen today in all horses and their relatives. The most useful examples of wild equids are probably the African zebra species, *Equus zebra, E. quagga,* and *E. grevyi.* Przewalski's horse, the true wild horse of Eastern Europe, has proven difficult to study under even seminatural conditions for a variety of reasons. Most experts believe it to be exterminated in the wild. There are several good studies of so-called wild horses and ponies in the United States and Great Britain, which we will mention later. (These animals are, in fact, managed to some extent, which does affect their ecology.)

Two distinct types of social organization are seen among equids, territorial and nonterritorial. The territorial pattern is less common, known to be utilized by Grevy's zebra, *E. grevyi,* and the wild ass, *E. africanus.* It has also been observed in populations of feral horses, *E. caballus,* under certain conditions.

The form of territoriality seen among Grevy's zebra is unusual for an ungulate. The territories are extremely large, each held by an adult male. Mares and their foals wander through adjacent territories at will; no close social bonds are formed with any other individuals. Other stallions are permitted in the territories as long as there is no estrous mare present. If such a mare is in the area, then the stallions do defend their boundaries. In effect, the mare chooses the stallion who will breed her by entering his territory. Outside of the breeding season, animals may form large, loosely associated groups or remain single. The only firm bonds seen are those between mares and their foals of the current year.

The nonterritorial, or harem, pattern is seen in domestic horses and most other equids. In this case each adult stallion maintains a family group of mares (or maybe even just one mare) and their offspring of up to about 2 to 4 years old. No territories are defended; a herd of plains zebras is made up of many family groups grazing in close proximity. The herd stallion keeps track of his mares and their foals, and is the only one who even attempts to breed the mares when they come into estrus. This provides a nice, safe, quiet environment for young foals, as there is none of the chasing and fighting usually associated with competition for mating.

Young stallions remain in their mothers' groups until they are about 4 years old. At this time they band together with other young stallions and older ones who do not have harems of their own. This bachelor group is very similar to those seen in various bovid species. When a stallion reaches maturity, at about 5 or 6 years of age, he begins to try to establish his own harem. No effort is made to take over another stallion's group, unless that stallion is obviously ill or infirm. The targets are the adolescent females, fillies of $1^1/_2$ to 2 years old. When they first come into estrus, they attract all the young bachelors in the area. At this time the herd stallion may try to drive them off, and some fighting often occurs. (See Fig. 4.) Eventually, some of the bachelors succeed in removing a filly from her family group, and continue to squabble over her among themselves. When she comes out of estrus, the filly usually returns to her family group again. This pattern repeats itself as soon as her next estrus begins. By the time she is 2 or $2^1/_2$ years old, old enough to conceive and bear a foal, the filly will stop returning to her family, and will remain with the bachelor stallion who succeeds in abducting her (Klingel, 1974; Tyler, 1972; Waring, 1983).

There is a real dichotomy in sex roles among the nonterritorial equids. The herd stallion is responsible for all movements of his group; he determines where, and how far, the group will travel each day. The quality of the resources available to mares and foals is, to some extent, up to him. Mares are responsible for only their foals. In a large herd, with many mares, a dominance hierarchy is usually established among the mares, but the stallion is still boss. Most herd stallions are very effective at keeping order in their families, including intervening in disputes between mares. If a herd member is

Figure 4   Fighting erupts between stallions when the young females in a herd reach sexual maturity. (Photo by B. Franke)

lost, the entire group will search for it, led by the stallion. Stallions drive lost or straying mares and foals back into their herd, even attempting to drag those that are too weak to walk (or that have been experimentally sedated; see Klingel, 1974).

Domestic horses, as we said, fall into the category of nonterritorial equids. Yet in most cases, mixed herds of mares and stallions are not maintained. The usual pattern on a stud farm is to keep the stallion in a paddock of his own. Mares are turned out in small groups together, the actual size of the group depending in part on how much pasture land is available. Mares and their young foals are usually pastured with other mares and their foals of about the same age. Yearlings of both sexes are put together, once the males have been gelded. Two year olds may be kept together as well, or integrated into older groups. Obviously there is little need for a gelding on a stud farm; these animals are sold for racing or riding. Fillies may be sold as well, or kept for breeding.

Although this system of grouping is not quite what we see in the wild, it is still quite satisfactory, as long as there is adequate grazing space. Horses are sociable creatures, and like contact with one another. Separation into peer groups (mares with foals, yearlings, barren mares, etc.) is appropriate. In the wild, each stallion has charge of a group of mares and their foals. Yearlings are not rejected by their newly preoccupied mothers, and turn to each other for companionship. Older colts, of course, form bachelor herds, and in some cases older fillies will do the same.

As long as the groupings remain fairly constant, there should be few problems. As with any social animal, stability is an important factor in a relationship. Peaceful coexistence among a group of mares depends on their familiarity with each other, and their awareness of their own status within the social order of the group. If new mares are constantly added, and old ones removed, this order never gets established. The resulting kicking and biting occurs as the mares attempt to find out who stands where. Constancy of grouping, as far as it can be accomplished, is of considerable importance to all our domestic species.

In most Western countries today, the horse is primarily a pleasure animal. There are still many places where it is a valuable beast of burden, but the majority of horse owners use their beasts for recreational riding, or producing foals. Whether it is to be a backyard pet or the next Secretariat, the production and maintenance of a healthy foal is an event of great importance to many mare owners. Proper medical care before, during, and after birth is essential, of course, and genetics will have a lot to do with how the foal finally turns out. But once again, good management of both mare and foal includes attention to behavioral factors.

A newborn foal is entirely dependent on its mother for protection. In the wild, the herd stallion would be an important defensive factor. Mares seem to be aware of their great responsibility, and become fiercely protective, especially in the first few days when the foal is most vulnerable. Both domestic and wild mares seek privacy when foaling is imminent; they never stray too far from the herd, but do try to isolate themselves from inquisitive noses. The birth process is amazingly fast, almost explosive, in horses. This minimizes the time that a mare spends lying down and helpless. Once the foal is on the ground, the mare turns to it and begins licking and nuzzling. This helps clear the foal's nostrils of clinging membranes, and is extremely important in establishing the maternal bond. (See Fig. 5.) Studies on goat mothers and kids have shown that early contact with the scent and taste of the offspring helps the mother to identify that kid as hers. Such

Figure 5   The first contact between a mare and her newly born foal is a critical step in the formation of the bond between them. (Photo courtesy of C. Uhlinger)

contact must occur very shortly after birth, while levels of the maternal hormone oxytocin are still high. The interplay of hormones and physical contact at the right time seems to be very important in "turning on mother love" (see Chapter 2). The same appears to be true of horses.

Thus it is very important that a mare be left undisturbed (as much as possible) with her newborn foal. Let her see it, and explore it with her nose and tongue; unless it is extremely cold, more harm than good will be done by rushing in to towel-dry the foal. In cases of difficult foaling, where medical intervention is required, the mare may remain down for some time. If this occurs, of course, someone other than the mare must clear the foal's nostrils and be sure it can breathe freely. If possible, the foal should then be moved up to the mare's head, so that she can smell it and perhaps even lick it. Often this contact with her foal will stimulate an exhausted mare to a miraculous recovery. When the mare is so ill that she cannot respond, the problems are rapidly compounded. The mare needs to receive the necessary medical attention, and the foal needs to nurse colostrum, the invaluable first milk that contains the antibodies it must have to combat disease until it can mount its own defenses. Any horse owner unlucky enough to find himself in this situation should seek the immediate aid of his veterinarian. Any vet called in will deal with the medical problems as they arise, but he should try to maintain as much contact as possible between the mare and her foal in this early period.

What happens if a foal is orphaned? In the wild, young nursing foals usually die soon after their mothers. Even if there are other mares with foals the same age, they will not permit the orphan to nurse. Unless there happens to be a mare in the herd who has just lost her own foal and is willing to foster the orphan, it will starve. Harsh as this sounds, it makes ecological sense. Unlike sheep and goats, horses usually have only one offspring. One is all they are able to provide milk for, at least under harsh environmental conditions. A mare who adopts an orphan foal does so at the risk of depriving her own offspring (see the section on kin selection); this just doesn't happen. Under managed conditions, however, there are exceptions. A well-fed Thoroughbred mare can produce more than enough milk for her own foal; we know of more than one case where a mare accepted an orphan foal of about the same age as her own baby and successfully raised both of them.

A common practice among breeders of valuable horses is to remove the foal from its natural mother shortly after birth and foster it on another mare who has recently foaled herself. (The foal of the foster mare, in turn, is raised on a bottle.) The success of this enterprise depends on the strong maternal nature of the foster mare as well as the age of the foal. The younger it is, and the closer to the birth of the foster mare's own foal, the more likely she is to accept it. It is usually necessary to restrain the mare for awhile, to allow her to adjust her responses. At first, the foster foal is seen as a stranger, an intruder, which stimulates an aggressive response. If the foal is permitted to nurse (while someone holds the mare), the mare will begin to see it as a baby. Finally, with a little luck, the mare's maternal responses will take over, and the foal will be accepted. Several points must be made clear here. First of all, adoption and fostering are not natural to horses, except under unusual conditions. Secondly, the behavior of the foal is very important in determining how the mare responds to it. An ill foal that does not act normally will not elicit the same caretaking behavior, in most cases, as a healthy, active foal. Wild mares will paw at a stillborn foal, or even a weak one, often causing considerable damage to its body. Finally, the choice of foster mother must be made very carefully. There are profound individual differences in behavior among mares, and this includes maternal behavior. The effects of domestication cannot be accurately measured, of course, but one of the traits we have selected for while breeding horses is motherliness. Whether she is to raise her own foal or someone else's, a mare willing to devote herself to the task is a valuable asset. (Readers interested in learning more about mare and foal relationships are encouraged to see the journal *Equus*, which contains many pertinent articles.)

Many of the common vices that perplex and annoy horse owners are related to feeding. The horse is different from most other farm animals in that it is being fed for athletic performance, not weight gain. In fact, knowledgeable horse owners are concerned with keeping their animal's weight down. Horses, as we discussed before, are designed to take in grass at a fairly constant rate all day long. On some farms, this is no problem. The animals are turned out to graze, and given supplements of grain or hay as they require it, but some stables work differently. For a variety of reasons (lack of pasture space, convenience, insect control, etc.) the horses are kept indoors in stalls, and only get outside when they are being ridden. These animals are not nutritionally deprived; they get high-quality, concentrate feed as well as hay. Yet for an animal whose ancestors have been grazing for centuries, it is rather an abnormal situation. Stalled horses often begin weaving back and forth, like the pacing of a caged tiger in the zoo. At the very least, they become bad tempered. (See Fig. 6.) Most of this behavior can be attributed to boredom; the horse is shut up in that stall with nothing to do (Kiley, 1974). Diversions such as hanging plastic jugs in the stall may help, as might a companion goat, or cat, or rooster. The best solution, though, is to try to give the horse as much chance to "graze" as possible. Put its hay in a rack, or a net, so that only small amounts can be pulled out at a time. Instead of two large feedings, give three or even four smaller ones. Obviously, this will make more work for the feeders, and may not fit in with the horse's workout schedule. But whenever possible, a stable manager should try to fit his routine to the natural behavior of his animals.

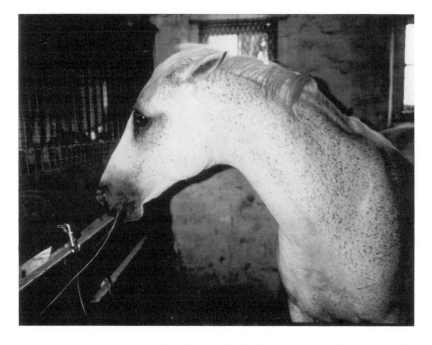

Figure 6 Horses confined to stalls for long periods of time may develop vices as a result of boredom. (Photo courtesy of C. Uhlinger)

The domestic pig, *Sus domesticus*, is another animal whose current existence bears little resemblance to that of its wild relatives, *S. scrofa*, the European wild pig, and *S. vittatus*, the Asian wild pig. Unlike the other ungulates, pigs did not develop into specialized feeders. In fact, they become so omnivorous as to allow them to spread over an extremely wide geographic range, throughout Europe, Asia, and Africa. There are many different species of wild or feral pigs today, ranging from small peccaries to imposing warthogs. Once again, differences in body type and behavior can be correlated to a significant degree with the habitat and resources available.

Most wild pigs are not truly territorial; they tend to be fairly sedentary, occupying one place for a long time, but they do not actively defend this area, nor do other pigs try to move into it. The exception is the warthog, *Phacochoerus*. This open-grassland dweller shows typical territorial behavior, defending its nest, sleeping, and feeding areas against intruders. Such behavior is not surprising. The savannah, unlike forested regions, offers little protection against predators. Warthogs thus remain in one place, with good burrows, as much as possible, and there is some competition for these spaces. Competition for resources is one of the precursors for the development of territorial behavior in a species.

This sort of competition does not seem to be an issue for the European wild pigs. These animals live in family groups, consisting of one or two females and their offspring, for about half the year. Adult boars are solitary, though occasional bachelor groups of young males are found. During the rutting season, the family groups split up as temporary mating pairs or groups are formed. Boars seek out the estrous females by scent, drive off the family group and other boars (or are driven off themselves by a stronger, more dominant male), and mate with the sows. A dominant adult boar may have as many as eight sows in his mating group at one time. He will breed with them frequently until they go out of estrus. At this point, the family groups gather again, and the boar goes off by himself. When parturition is near, the pregnant sow either leaves her now nearly grown offspring or drives them off. She builds an elaborate nest, gives birth to her new litter, and, when they are weaned, may join with another female to form a new family group. The abandoned adolescents usually split up, the males going off on their own and the females, when they come into estrus, following their mother's pattern. Occasionally, young females will remain with their mother for another year (Pond and Houpt, 1970; Hafez, 1975).

This pattern of young males leaving and young females staying should be looking familiar by now; it is common among social animals. It is an excellent way to maintain a stable social group without too much inbreeding. The males leave, so there is no risk of their mating with their mothers. There is a chance that a young male will mate with his sister, if she leaves too, but it is a small chance. It is unusual for a male to retain dominant status for too many years, which minimizes the number of father–daughter matings. A few such crosses are no problem, and in most cases the breeding male will have been replaced before there is chance of grand-father–granddaughter crosses. The young males, when they leave their family groups, tend to go some distance away. Often they have to travel in order to find a place for them-selves. The actual distances vary between species, of course, but in general males disperse more than females. This has the effect of enlarging the gene pool even more, bringing new blood into the family where the male eventually breeds. There are a few species where the females disperse, but this is very rare.

Several years ago, even the largest hog producers kept their animals in outdoor pens. Now more and more are turn-ing to total-confinement operations, where the animals spend their lives indoors, moving from one house to another as they mature. Except for breeding stock, these lives are pretty short, with most going to market at about 6 months of age. Animals are separated into groups according to their age, size, and food requirements; a piglet may not remain with his siblings if he is considerably smaller or larger than they. Mixing litters so that each sow has an equal number of approximately same-sized pigs is a common practice as well. Adult sows are usually kept penned in groups of four or five during the periods when they are not nursing young; boars are kept separately, in individual pens. This arrangement, at least, is not too different from the grouping seen in the wild, but the entire structure of the confinement operation is so unlike anything found in nature that it is impossible to make very many comparisons. Fortunately for producers, all the hogs have to do is eat, and they usually do this very well. Behavioral problems seen on large swine farms are more often related to overcrowding; the animals fight and bite. This can be easily dealt with by providing more room, reducing group size or reorganizing groups if it looks as if one or two bullies are picking on the smaller, weaker pigs. Other problems, such as boredom and its accompanying vices, are rarely seen because of the shortened life span of these animals.

Some pigs are raised outdoors in the dirt, of course. (See Fig. 7.) Though this practice is avoided by major producers in much of the United States because of parasite problems, it is still the easiest way to raise pigs on a small scale. There is no doubt that a sow blissfully wallowing in mud with her litter appears more contented than one confined to a crate where she cannot turn around. Domestic pigs left to root and forage on their own will form social groups just like wild pigs. The sow makes a den with a nest for her litter, and the young pigs

Figure 7   Pigs are sociable animals, and thrive together in small groups. (Photo by G. Honoré)

remain in it for a week or two. When they emerge, there is much playing, running, and tussling. The scene is reminiscent of a litter of puppies.

Although the major function of the domestic pig in the world today is to produce pork, it is used in a variety of other fashions. Some people find that pigs make excellent pets. Who can forget *Charlotte's Web*? A British sow named Slut was trained by her owners to point game. She would come running at the sight of the gun, eager as any hunting dog. French pigs have been used for centuries to sniff out truffles. Pigs' olfactory senses are highly developed, and they can be trained to recognize specific scents and respond to them. Perhaps they have a future at airports, sniffing for explosives in luggage. Pigs are credited with high intelligence. While ranking animals according to intelligence is a difficult (and questionable) task, there is no doubt that pigs learn quickly and are readily trainable. This attribute has made them valuable research subjects for psychologists. Their size, reproductive habits, and the fact that they can be maintained in close confinement are making them more and more attractive to researchers in other fields. Pigs are physiologically more similar to humans than are other nonprimate animals, and are therefore very useful in biomedical research. In fact, valves from pig hearts are often used as replacements in failed human hearts. Pigs may soon be one of the more popular laboratory animals, as well as a desirable and economical source of meat.

In management of a swine herd, most vets will be kept busy with disease prevention, however, the issue of overcrowding is also important, as with cattle. Maximizing production means minimizing stress, and that means keeping an eye on group size and composition. Young piglets are commonly grouped by size in the nursery. This is an acceptable practice, but the groups should not be re-sorted as the pigs grow. Although re-sorting maintains parity of size and weight within groups, it means the disruption of established social orders, added stress, and an increase in the likelihood of injury or disease. Remember that pigs are intelligent, olfactorily oriented animals when you plan your operation. Social contacts are essential, but so is room to escape from others. If there are libido problems with the breeding boar, be sure he can smell the female, even if artificial insemination is the method being used. Perhaps swine-raising practices have evolved faster than swine; the latter still retain many behavioral similarities to their wild relatives.

C. Carnivores: Dogs and Cats

The domestic dog, *Canis familiaris*, and the domestic cat, *Felis catus*, both belong to the group called Carnivora, the flesh eaters. Dogs are members of the family Canidae, which includes wolves, coyotes, and jackals, among others. Cats are of the family Felidae, along with lions, tigers, cheetahs, and various smaller relatives. Although there are some apparent physical similarities between dogs and cats, a result of their common food source, there are even more differences, both physiological and behavioral. We know an anatomy professor who reminds her students at the beginning of the semester that "a cat is not a small dog." The more you know about the two, the more apparent the truth of this remark.

There are more varieties of domestic dog than of any other domestic species. (See Fig. 8.) These variations may be striking. When we were discussing ruminants, we mentioned that it seems impossible that the giraffe and the duiker are in the same family. How about the St. Bernard and the Chihuahua? They are not only in the same family, but the same genus, even the same species. If physically assisted, they could breed and produce offspring. Many of our modern breeds are the result of generations of careful selection for certain characteristics. Others have existed for centuries just as we now find them. Behavioral characteristics vary as well as physical ones, of course. "Everyone knows" that Dobermans are "more aggressive" than Beagles. The quotation marks are present because aggression is very difficult to define, let alone quantify. Yet this statement is accepted as common knowledge by most people even peripherally involved with dogs (except maybe some Doberman breeders). Despite these variations, there are some behavioral traits that may be said to apply to all dogs, regardless of breed. In fact, within the

Figure 8 Compare the physical characteristics of these bulldog puppies with those of a German Shepherd or a poodle! (Photo courtesy of J. Armstrong)

Canidae, one finds relatively stable rules of social organization. There is less behavioral specialization between species than is seen in other groups of animals.

One of the better-studied wild relatives of our domestic dog is the wolf, *C. lupus*. Many geneticists believe that the wolf is one of the dog's closest relatives as well. Certainly some breeds of dog are very "wolfy" in their appearance; take a good look at a Malamute sometime. In other breeds the physical resemblance is less strong, but the dog has been domesticated for thousands of years, and selective breeding for certain traits is a powerful shaping tool.

Canids, whether wild, feral, or domestic, are social animals. They live in groups, or packs, and show varying degrees of mutual cooperation and interdependence. Most food is obtained by hunting prey, though all canids will scavenge to some extent. There are three main patterns of hunting: solitary, transitional, and social. Solitary hunting will be mentioned again when we talk about cats. This is the most common method employed by felids, and is seen much less commonly among canids. A good deal of stealth and patience are required as the hunter sneaks up on the unsuspecting prey. Wolves practice a variation on this theme, whereby one member of the pack hides and sneaks while the others drive the prey toward him. Many of the small canids, like the foxes, are also solitary hunters. The pattern of hunting depends largely on the size of the hunter and the size of the prey. Foxes eat small prey relative to their own size, and are also among the most omnivorous of the canids. This lifestyle ideally suits them for solitary hunting.

When the prey gets larger, the predator needs help. One coyote can catch and eat a mouse all by herself, but it may take a pair of coyotes to run down a jackrabbit. Hunting in pairs, especially when the pair bond is fairly permanent, is representative of the transitional pattern, not quite solitary, but not truly social. Animals utilizing this pattern usually rely on solitary hunting as well, for smaller or more densely located prey. Cooperative hunting might have evolved in response to a sudden decrease in the usual food supply. No more mice? Well, what about rabbits? But rabbits are bigger and faster than mice, and take a lot more time and energy to catch. They also provide more meat, of course; enough for two. Since only two animals are involved, or at the most a pair and their nearly grown young, it is important that the solitary skills are not lost altogether. If something happens to the hunting partner, the remaining animal must be able to carry on alone (Fox, 1975).

True specialization of hunting skills is seen in the social hunters, where the whole pack hunts as a unit. Of course, a wolf is also able to hunt small prey by himself, or with his mate. But the most social canids, those that live in relatively large packs with complex dominance relations, have developed a means of preying on even very large animals such as moose and caribou. No lone wolf has a chance of bringing down one of these beasts, unless it is nearly dead from some other cause. Once again, the availability of prey is a determining factor in the size of the pack. Very large packs of up to 30 or 40 wolves were reported on the midwestern prairies when these lands were densely populated with bison. In regions of Canada today, where the elk are spread more sparsely, packs of 8 to 10 wolves are common. There is no point in getting together in a big group to bring down prey if there will not be enough meat to go around.

Learning to be a skilled hunter, whether solitary, social or in between, is vital to the survival of most wild canids. Obviously, this is not the case for the domestic dog, at least when it is kept as a pet or research animal. There are societies where the dog is intentionally left to forage on his own, and many other cases in our own society where pet dogs are turned into feral nuisances by irresponsible owners. A properly cared for domestic dog has no need to hunt for his dinner; yet generations of hunters lurk in his genetic past. Is there truly a "genetic urge" to hunt and kill, or are these just skills taught to pups by their parents? The answer lies somewhere in between those two statements. (Refer to Chapter 2, Section D on instinct and learning.) The process of socialization is extremely important to the development of both the wolf cub and the domestic puppy. The more rules, skills, and traditions there are in a society, the more important it becomes to have a way of learning them, to ensure the survival of both the individual and the group. This is obviously true in human culture, and appears to apply to animals as well.

In 1965, Scott and Fuller proposed that the development of a puppy be divided into four stages: neonatal, transitional, socialization, and juvenile. This division has remained popular and useful for years, with minor revisions. Like any set of stages, it is only a guideline, a framework for understanding the linkages between physical and psychological development.

The neonatal stage covers the period from birth to about 2 weeks of age. During this time the puppy is physically helpless, and its eyes and ears remain closed. It will respond to tactile stimulation from the mother (or other sources) with reflex actions only. The nervous system of a young puppy is still not fully developed. Both motor skills and the ability to receive and process sensory input are minimal. Thus for the first two weeks of its life, the puppy is suited only for life in a safe nest, with mother to provide food, warmth, and stimulation to urinate and defecate.

During the transitional period, roughly the last half of the first month after birth, rapid maturation of the nervous system occurs. Full development is not complete, but the puppy's capacity to receive and respond to stimuli of various sorts is greatly increased. Motor skills also improve radically. The pups now begin to crawl, to walk, to tumble about. Eyes and ears open, and it is suddenly very difficult to keep the pups in the nest. Distress vocalizations begin at about 3 weeks. This is one of the first occasions when the puppy actively demands attention and care from its mother. Up to this point, the mother has responded to reflex motions by her pups (such as rooting at the nipple) or acted on her own initiative. Now the pups begin to take an active and manipulative role in soliciting care, the first step in the socialization process. Because of its immature nervous system, the month-old puppy is still not very trainable. Studies have shown that pups of this age are able to make discriminatory responses, and demonstrate avoidance learning, but they are not capable of much more. Those skills that develop first are the ones essential for survival. Even a very young puppy will cry and squirm when placed on a cold surface, or when removed from the comforting pile of littermates. Slightly older pups follow mother about the nest site as best they can, and cry for her attention. Other categories of behavior, such as play or exploration, can wait.

The period of socialization, from about 4 weeks to 10 weeks, is one of the most significant times of a puppy's life. (See Fig. 9.) Maturation of the nervous system progresses, motor skills improve, and the puppy is ready to plunge into the business of learning to be a successful member of society. This means learning to fit into the pack, to communicate with other dogs, as well as to fit into human society if required. The concept of imprinting was discussed in an earlier chapter. Though few people now believe that socialization in dogs is an all-or-none thing, that if it does not happen at a specific time, it will not happen at all, there are times in a dog's life when it is simply more receptive to certain stimuli than it ever will be again. Once again, this makes sense from an evolutionary point of view. By this time, the puppy is physically mature enough to be mobile, active, and aware of its surroundings. It no longer needs to devote all of its time to feeding and staying warm. On the other hand, mother is still

Figure 9 During the period of socialization, puppies are most receptive to learning the rules of canine and human society. (Photo by P. Klopfer)

nearby, still the provider of food and the protector. The pups are still small and immature in appearance, so their blundering behavior does not elicit the same aggressive response from other adults in the pack (or neighborhood) that a socially inept adult would. Even dominant males in a pack tend to be rather tolerant of boisterous puppies. This tolerance decreases as the pups grow, until eventually they are treated like the other dogs in the group. This period of relative freedom to make social errors without punishment gives the puppy an opportunity to learn how things are done. Inappropriate behavior is pointed out, of course, but with a quiet growl or show of teeth rather than a full attack. During this time the pups will also get a chance to make social contacts with other litters, and unrelated males. Playing, which serves an important function for dogs of all ages, is an integral part of the development of social skills and relationship (Fagen, 1981).

Needless to say, this familiarity is essential to the proper functioning of pack society, be it a hunting group of feral dogs or the neighborhood gang.

From the point of view of the pet owner, the period of socialization is also of great significance. By the age of 6 or 8 weeks, the puppies are physically ready to leave mother; they can be weaned to solid food, and are able to thermoregulate on their own. Moreover, as we have just discussed, they are ready to expand their horizons, to meet new people (canine or human) and to adapt to new environments. This is the ideal time to bring a new puppy home. (See Fig. 10.) It will be at its maximal point of receptivity for socialization, and the adjustment to a new family (again, canine or human) will be easiest and most complete. Playful behavior is also important in establishing relationships with people. A younger pup is probably not yet ready for this change. Even though you can raise it on a bottle, or even wean it early, it is psychologically unprepared for the challenge of becoming a member of a new group. Such puppies often develop nervous habits, or retain their old reflex motions into adulthood. Frequently this takes the form of sucking, either on the dog's own paws or tail, or on a piece of cloth. The appearance of infantile behavior usually signals that the animal is feeling insecure in some way, and is seeking the comfort it used to find with its mother.

Should the older puppy, 11 weeks or more, be even more ready for integration into a new household than it was at 7 weeks? In general, no. Although we said that the concept of imprinting could not be strictly applied to dogs, it is true that there are sensitive periods, finite times during which the animal is especially ready to react to certain situations. The sensitive period for socialization in dogs appears to be roughly between 6 and 9 weeks. Obviously, there is no window in the pup's brain that slams shut at 10 weeks, allowing no further input. No one is sure quite how the process takes place. We have discussed the evolutionary reasons for a grace period in learning social skills, and the fact that this period is coordinated with the maturation of the nervous system and the physical growth of the pup. Beyond 10 weeks, this time of trial and error fades gradually into normal adult life. The pup learns its role and begins to fit into its place in society. This

Figure 10   Eight weeks old, this puppy is rapidly learning its role as a dog, but its social blunders are still tolerated by adults. (Photo by G. Honoré)

time is referred to as the juvenile period, between the stage of rapid maturation and acceptance of new situations, and sexual maturity and the full assumption of adult roles. The significance of this development to us is that now it becomes more and more set in its ways and less receptive to change. When you remove a 12-week-old pup from its family group, or even from its solitary kennel, you are disrupting its life a lot more than you would have 4 weeks earlier. Of course, the puppy will adapt to its new life in time, but it will probably always be a bit more reserved, less devoted to you and your family than if it had been adopted earlier. This is because, during its period of maximal receptivity to new social stimuli, it was becoming socialized into a different group.

If you do bring home an older pup or even an adult dog, do not think your chances of a happy relationship are over. Dogs are by nature very sociable and adaptable creatures, it just takes an older one a little longer to settle into a new society. Veterinarians are often consulted about behavioral problems in clients' dogs. This is too large and complex a topic to cover here, but in many cases these problems arise from improper socialization. Either the dog feels constantly threatened, and responds with fearful snapping, or the owner has not managed to properly establish her own dominance, and the dog keeps challenging it. Vets should be on the lookout for these signs, and encourage all dog owners to do at least a minimal amount of obedience training with their dogs. After all, canine society has many rules of its own, and dogs are designed to learn to cooperate and live peacefully in a group situation.

Cats are another story! The lion, *E. leo*, is really the only felid that has developed a cooperative society. Although lion behavior is fascinating, it is so different from the lifestyle of most other felines, including the domestic cat, that we will not discuss it here. (Anyone looking for prime examples of male chauvinism in action should certainly take a look at lions, though.) The rest of the felids tend to be more or less solitary. Some live in pairs for at least part of the year, and kits often remain with the mother until they reach maturity. Territoriality is seen in many species, while others just seem to avoid one another's home ranges.

Felids are found all over the world, adapted to all kinds of environment and terrain. As we mentioned earlier, they are (with few exceptions) solitary hunters, relying on stealth and concealment to capture their prey. Recall that noncooperative hunting means smaller prey. Many good-sized cats still feed on small birds, rodents, and reptiles. Others have found different ways of dealing with the food problem. Canids manage to take advantage of larger prey by hunting as a cooperative unit, thus increasing the predator mass. Felids handle the problem by increasing the individual predator mass. Thus we have tigers, several hundred pounds heavier than the largest canid, able to catch and eat big prey all by themselves. These different evolutionary strategies for successful exploitation of potential food sources have resulted in very different behavioral strategies as well. Dogs need to cooperate with one another, and their society is directed toward developing this skill. Cats are very much more independent. In fact, their success as hunters depends on the absence of other cats in the vicinity. This trait is evident in domestic cats as well as in many wild species.

The African cheetah, *Acinonyx jubatus*, is known as the fastest mammal on earth. This famous speed is but another adaptation to solitary hunting. Cheetah social structure bears some resemblance to that of domestic cats, and is probably one of the best representative models of a felid society. Adult cheetahs are usually found alone. Females may be seen with their litters, and males sometimes form loose associations similar to the bachelor herds seen in some hoofed mammals. These groups are often made up of brothers, from the same litter, who roam off together when they are about a year old. Though both male and female cubs leave the mother upon reaching maturity, they will occasionally encounter her again. Researchers have reported seeing groups of three or four adult cheetahs plus the current litter of cubs. In at least some of these cases, the additional adults have been identified as older offspring of the mother of the cubs. Thus even solitary animals are not entirely asocial. A female with cubs tends to avoid association with strange cheetahs of either sex. This is relatively easy to do, as the strangers in turn are trying to avoid her. She will become territorial for the period in which she raises the cubs, defending the area around her den with considerable intensity. For the rest of the year, we should probably refer to her home range rather than her territory.

Some cheetahs, male and female, become nomadic. They travel in a specific direction, crossing other animals' home ranges as they go. These individuals do not make boundary markings, nor do they pay any attention to the marks of others. Most other cheetahs do not seem to mind their passing through, so long as they keep going, but territorial males object violently. The rare fights between adult cheetahs usually occur when a nomad strolls across a border into the territory of a male who takes his boundaries seriously (Eaton, 1974; Wrogenan, 1975).

We described the determinants of territoriality when we discussed ruminants, and we can apply the same guidelines here. When food (and females) are scarce and scattered in distribution, it is to a male's advantage to defend a territory. When resources are more abundant, it is probably not worth the time and energy it takes to keep others out, and home ranges only are maintained. Since the cheetah's lifestyle is not based on cooperation, and in fact depends on a degree of hunting stealth that can only be practiced alone, the animals naturally tend to spread themselves out over a large area.

Scent marking is an extremely important means of communication among all felids, because it leaves a lasting signal; the sender need not be present when the receiver gets the message. A scent mark (urine, feces, or glandular secretions) can provide information about the age, sex, physiological status, and individual identity of the sender, as well as the approximate time when marking occurred. This is how cheetahs (and most other cats) are able to maintain their spacing. Rather than having direct physical encounters with one another and communication through visual or auditory signals (although this happens as well), they keep track of events through this neighborhood bulletin board.

Domestic cats, when allowed to run freely outdoors, have a similar bulletin board arrangement. Studies on feral cats in European cities show that each individual maintains its home range, which varies extensively in size. Some cats are even territorial. A network of safe paths is formed, connecting various important areas, such as feeding sites (garbage cans), water sources, and a group assembly area. These paths either avoid crossing individual territories, or are tolerated by the resident. Similarly, the assembly area is a neutral space, where all the cats in the area can gather for any reason. Often these areas are used for courtship. Anyone who has ever walked through the back streets of a large city has undoubtedly seen such a congregation of cats, just sitting around. Some social encounters occur; cats enjoy mutual grooming, especially between females and their young, but also between two animals that know each other well. Cats do recognize one another, and there are dominance hierarchies to be respected. Often these hierarchies do not cross sexual lines; males dominate other males, females dominate other females, and the two sexes have little to do with one another unless the female is in estrus. A strange cat that suddenly shows up in the neighborhood is treated with great suspicion, and immediately challenged by the residents, especially those of its own sex. Unless it shows prompt submission to the threat displays, a screaming fight will ensue.

In most encounters between individuals, true fighting is rare. (See Fig. 11.) Two cats meet on a path. They stare at one another, recognition occurs, and the less dominant of the two moves out of the way. Cats have a variety of stereotyped moves to indicate their status and their intentions. We have all seen two cats crouching, immobile but for a possible tail twitch, glaring at each other with unmoving eyes. Every muscle is taut, and we wait for the leap, the screech, the flying fur. Suddenly, one of the two breaks the stare, turns away and begins washing himself vigorously. The other may maintain position for a moment or so, but then gets up and stalks off. A battle has been won without ever taking place. We obviously cannot tell what each cat saw in the other's face, but the message is clear. Cats use their whole bodies to signal one another. We can see all these movements, taken out of context, in kittens at play. While still young, they practice the repertoire of motions that will be required of them as adults. Gradually they learn to use them in appropriate situations, and visual communication takes place (Leyhausen, 1979).

Figure 11   Cats are highly olfactory animals, and often use a system of scent marks to communicate with one another. (Photo by G. Honoré)

Olfactory (scent) communication is also very important to domestic cats. This is a frequent source of dismay and frustration to the pet owner. (See Fig. 12.) A tomcat does not spray urine to be contrary; he is signaling to other males that this is his territory and they had better keep out, and to the females that he is here and virile. It is impossible to explain to a cat, especially an indoor cat, that there is no need for this sort of communication. Some males can be trained not to spray in the house, but it is a very difficult task. Often the worst offenders are those males who have had an opportunity to interact with other males, who are fairly dominant, and who are now confined to a house and see other cats walking down the street. The frustration must be like desperately wanting to say something and being unable to speak or be heard. This is a difficult situation to deal with, as the need to communicate is clearly very strong in the cat. Punishment usually does not work too well. The best solution, in the long run, is to neuter all male cats that are not valuable breeding studs. Not only will this reduce the desire to spray (when the cat no longer feels hormon-

Figure 12   While an entirely normal part of feline social behavior, urine spraying by males is not desirable in a house pet. (Photo courtesy of J. Armstrong)

ally male), but it will also help curb the rampant overpopulation of cats. Unfortunately, spraying has sometimes become such a habit that castration does not help. Veterinarians should be aware of the potential for this problem, and counsel all owners of male house cats to have them neutered as soon as possible. (Our personal feeling is that owners should be counseled to have all cats and dogs, male or females, indoor or outdoor, neutered as soon as possible. There are enough unwanted kittens and puppies in the world as it is.)

Just as with dogs, the list of behavioral problems that arise between cat and owner is too long to deal with here. Bear in mind that many of these problems are related to the cat's special evolutionary quirks. Physiologically, cats are more purely carnivorous than dogs. Their protein requirements are much higher, and less easily met by vegetable substitutes. They are unable to synthesize certain amino acids, and must get them from the meat they eat. Behaviorally, as we have seen, they tend toward a solitary lifestyle. This does not mean that they do not enjoy the company of other cats (or people), but do not expect pack behavior from a cat. Finally, they are highly olfactorily sensitive animals. Much has been made of the dog's sense of smell, but the fact remains that dogs are generally interested in, not repelled by, what they smell. A cat can be disgusted by a dirty litter box and turn to using the carpet instead. If the carpet is then cleaned, the remaining faint smell is just enough to remind the cat, or signal others, that this is the place to urinate. The problem perpetuates itself.

Cats are more difficult to train than dogs, partly because their less-social lifestyle does not require them to learn so many rules. Dogs are often described as eager to please; no one would ever say that of a cat. This does not mean that dogs are smarter than cats, or vice versa. It merely indicates the great differences in their social structure. Learning to be a successful dog entails learning to cooperate with others; this propensity can be put to good use by humans. Dogs often seem to consider their people as members of the pack, which can then result in both the submissive and the dominant responses we have discussed. Cats do not have packs. For all we know, our cats may consider us members of the neighborhood, to be interacted with at the group assemblies. But there is no call for cooperation in cat society, hence no eagerness to please. (See "The Cat that Walked by Himself," Kipling, 1902.) We can get much more pleasure from our pets (and they from us) if we try to remember their evolutionary roots and not expect the unreasonable. Deal with a dog in dog terms; communicate with a cat in cat terms.

D. Avian Flocks: Chickens, Turkeys, Ducks, and Geese

This section concerns those birds commonly raised for food production, either meat or eggs. Some are of great commercial importance and are extensively managed, such as laying hens in "egg factories," while others are mainly found in backyard flocks. The effects of domestication vary from one type of bird to another, depending on how much artificial selection (breeding and hybridization) for specific traits has taken place. Broiler turkeys are very different from their wild ancestors in many ways, while Toulouse geese resemble the wild Greylag in more than just physical characteristics.

Although the four groups of birds listed in the title are not closely related to one another (chickens and turkeys are in the family Gallinacae, ducks and geese in the Anatidae), some generalizations may be made about them. These generalizations apply also to certain managed game birds, such as grouse and quail, which we will not directly discuss here.

First, all these birds have *precocial* young. This means that the hatchlings emerge from the egg covered with down and able to move about and forage for themselves in a matter of hours. Contrast this with species having *altricial* young, such as the songbirds, where the hatchlings emerge naked and helpless, and may take weeks to mature enough to leave the nest. In addition, these birds are all flocking species (though wild turkeys in woodlands are often solitary). They tend to group together in avoiding predation, as we shall see. Group living requires a degree of cooperation and communication not necessary for a solitary life; recall the comparisons of canid and felid behavior. Thus, the young of these birds must learn a set of social guidelines. Finally, the concept of imprinting has great significance for all precocial birds, as in the famous studies by Konrad Lorenz with goslings. Other investigators of imprinting have used chicks, ducklings, turkey poults, and hatchling quail. These birds are the model for the whole theory. Appropriate social behavior must be acquired by all animals if they are to survive and propagate their genes. The method of acquisition varies considerably, of course, depending on the complexity of the society, the life span of the animal, and the type of parental care provided.

Altricial birds remain in the nest for weeks, dependent upon parental feeding. During this time, they can hear the calls of their parents, other birds of the same species, and different birds and animals. Later they can see what is going on around them, without taking an active part. This passive period (the only activity is begging for food, which can take up much of the day) allows the young bird to become familiar with its surroundings before it has to begin interactions with the world.

Precocial species are out in the world as soon as they shed the eggshell. The parents are not responsible for providing food, although many species will lead the brood to a likely spot and demonstrate pecking at worms or seeds. With a large brood, it becomes impossible for the mother to keep track of everyone. Much of the responsibility for staying with mother rests on the hatchling. Thus, the significance of the following response. Evolutionary demands placed on precocial species have led to young birds' being susceptible to certain stimuli at certain times. In this fashion, their safety and survival are assured, and so is their integration into society. Early in life, a duckling "learns" to recognize and follow its mother. Later, it "learns" what a duck of the opposite sex looks like. We use quotation marks because this is not learning in the classical sense; at appropriate times, the duckling seems able to absorb the information it needs from its surroundings. Learning becomes important once the brain has been aroused by the stimulus. Lack of the right stimuli at the right time may lead to problems, both immediate and in later life. (Refer to the Chapter 2 Section E ii on imprinting.)

The domestic chicken is known as *Gallus gallus* or *G. domesticus*, depending upon whether you believe it has one ancestor or many. Some researchers believe that the Red Jungle Fowl, *G. gallus*, is the single ancestral species, and that all other breeds and races of chicken stem from it. Others feel that there were several ancestral species, and that the modern domestic chicken in all its variety is the hybrid result, a new species. At any rate, everyone agrees that the Red Jungle Fowl is an important ancestor (if not the only one), so it serves as a useful model for wild chicken behavior.

Jungle fowl aggregate into large flocks of adults and juveniles. These flocks forage together, and roost communally at night. The mating system is polygamous; no pair bonds are formed. A rooster will mate with as many hens as he can, and makes no effort to defend a particular hen against the advances of another male. Preferences for certain hens do exist, however. Hens respond to the sexual advances of any male, though the intensity of the response is directly related to the intensity of the male's courtship. Again, it has been shown that certain hens also develop preferences for certain roosters, so the mating pattern is not entirely random. Both sexes appear to be involved in making choices (Wood-Gush, 1971).

Male Jungle Fowl are extremely pugnacious. Many a farmer has been taken by surprise as one of these tiny bundles of feathers and spurs flies in his face with a crow of triumph. Despite this fact, there is relatively little fighting among the members of even a large flock. Of course, the term pecking order originated with chickens. It refers to the dominance hierarchy which emerges in a group in which everyone knows everyone else. This hierarchy may be strictly linear, or triangular (see Chapter 5, Section A on Ruminant Herds). Like all dominance relationships, its success in maintaining peaceful coexistence depends on familiarity. If a new bird appears, fights will erupt continually until his place in society is determined. Visual signals are very important to most birds, including Jungle Fowl, and the posture and gestures of individuals contain a wealth of information. Submissive birds will crouch down silently and sleek their feathers close to their bodies when approached by a more dominant flock member. Eye contact is avoided, if possible, and the beak points downward. The whole bird looks smaller, and any movements that might conceivably appear threatening are absent. The dominant bird, on the other hand, advertises his top-dog status to the world. His feathers, especially those on his ruff, are fluffed to their fullest. He stands tall on his tiptoes, flaps his wings and crows. He may approach the crouching subordinate with a series of hops and feints, reminiscent of fighting moves, all the while directing his gaze pointedly at the other bird. In short, he tries to appear as large and aggressive as possible. While the other male remains in submissive posture, peace ensues. The dominant bird has made his point, and will soon return to his normal size and shape and begin scratching for food. Roosters often take this approach to hens as well to subordinate males. A receptive hen will respond to such a supermale display with a stereotyped display of her own, indicating her willingness to mate. If she does not, the male will probably try out his act on another hen. Crossmatings between certain breeds of chicken are entirely unsuccessful, simply because the female fails to respond to the male's display, or vice versa. Proper communication, in this case by visual cues, is essential.

When we talk about the behavior of domestic chickens, we run into the same problems as with domestic swine. Modern poultry operations are geared toward the most effective production of a commodity (meat or eggs). The bird's lifespan is artificially short, as is the case with any food animal. In addition, the processes of domestication and selective breeding have led to significant changes in both the morphology and the behavior of at least certain breeds and strains of chickens. The behavior of a backyard flock of a not-too-tampered-with, all-purpose breed of chicken is very similar to that described for the Red Jungle Fowl. Most domestic roosters are less aggressive then Jungle Fowl, but the same patterns of courtship, polygamous mating, and brooding of precocial young are seen.

In commercial breeds of laying hens the tendency toward production of eggs is the trait that has been selected for. These breeds, or strains, lay far more frequently than wild or all-purpose chickens. Ordinarily, a hen continues to lay until something signals her to stop making eggs and start brooding those in the nest. A broody hen sticks tightly to her nest, keeping the eggs warm and protected. This behavior is vital to proper incubation and a successful hatch. In commercial laying hens, broodiness is undesirable. The hen is not supposed to stop laying. Thus we see selection for nonbroodiness, with the result that the trait is nearly absent. Hens of layer stock make terrible mothers, even when put into naturalistic surroundings. They do not make nests, but drop their eggs all over the ground. If the eggs are gathered into a nest and the hen is placed on top of them, she will not brood them. This loss of broodiness is often used as one of the classic examples of the effects of domestication. It is more accurately an example of the effects of selective breeding for a certain trait, since the disappearance of the behavior was not incidental to another process (Hafez, 1975; Craig, 1981).

In broiler chickens, on the other hand, we see the effects of domestication. Here the bird has been bred for generations to produce a large, meaty carcass. Its behavioral characteristics are of no concern, except for the urge to peck and feed. Thus even the most highly selected strains of broilers retain fairly normal behavior. There are variations from strain to strain, of course, as one would expect in any highly inbred line. Geneticists find psychological peculiarities are linked with different lines of chickens, just as they are with rats. In general, if a broiler hen is put into the same situation as described for our layer hen, she will brood and raise the young. This is not always the case. Any time the genetic makeup of a bird (or animal) is altered to cause a morphological change, there is the chance that there will be behavioral changes as well. These are not always predictable, although in an inbred strain that has stabilized for several generations, predictions begin to have some meaning.

The main behavioral problems that plague chicken producers are seen in broilers, for the simple fact that they are kept loose on the floor, while layers are usually confined. In the old-style laying houses, where the hens are also left to run loose, the same problems arise. Chickens tend to peck. This is a highly useful behavior in young chicks, because their inclination to peck at every spot they see eventually leads to feeding. Unfortunately, they also peck at one another's eyes, nostrils, and any other specks. Thus a small wound on one bird can lead to veritable cannibalization as its companions peck at it. The modern practice of debeaking, removal of part of the upper mandible, was developed in an effort to curb this problem. The birds can still feed, but do not inflict so much damage on one another.

There is really very little behavioral management in the chicken house. Crowding is not usually a problem, except in extreme cases. A good manager will note whether or not all individuals are getting a chance to eat and drink; often the subordinate birds are kept away from feeders and waterers, so then more sources should be provided. The same applies to heat lamps with young chicks. They like to pile up together, but a massive pile under one lamp may suggest that more lamps are needed. Once again, try to keep social groupings constant. All those chickens may look alike to you, but they do recognize one another, and strangers will be picked on. It is a good idea to group the birds roughly by size and age. Not only is this socially easier for them, but it also reduces disease transmission and makes it easier to move the birds to market when the time comes.

In the case of laying hens that are allowed to run loose in a house, be sure there are adequate nest sites, or at least enough litter on the floor that the eggs do not get broken. If roosters are kept with the hens, try to keep the minimal possible number. Fighting may occur if there are too many males in a confined space. Finally, if you have backyard chickens, you may get more pleasure than you think out of watching them. A lot goes on in a chicken flock.

The domestic turkey, *Meleagris gallopavo*, is a direct descendent of the wild turkey bearing the same scientific name. There the similarities may appear to end. The American wild turkey was proposed as our national bird. Ben Franklin praised it for its intelligence, its cunning, and other fine qualities. Yet domestic turkeys are often described by poultry producers as the stupidest birds around. This is not quite fair. Much of what makes turkeys appear stupid is simply behavior that would make sense in another context. Poults will peck at electrical cords and kill themselves, but after all, curiosity and the urge to peck at things are traits that help young turkeys to survive in the wild. Electrical cords have not been around long enough for selection against this behavior to take place. Turkeys have been known to drown themselves trying to catch falling raindrops, but these birds eat a lot of flying insects, and coming out of the rain appears to be a habit learned from mother. A group of poults with no parent present may be forgiven for mistaking raindrops for bugs. We do not want to make too many excuses for turkey behavior (our respect for the common sense of turkeys is rather low); we just want to caution, once again, about the many ways in which an action may be interpreted.

Five different subspecies of wild turkey are found in the United States, from the wetland swamps of Florida to the Ohio woods, and even the scrubby brushland of New Mexico. Slight variations in social behavior appear with variations in habitat, as we might expect, but the general pattern remains the same.

Although they are in the same family as chickens, turkeys behave very differently. Large, mixed-sex flocks form only at night when the birds roost. Even these roosts are different. Chickens pile into one or two trees together, loading the branches down as they crowd next to each other. Turkeys all roost in the same area, but each bird occupies its own branch, or even its own tree. Definite preferences for roost sites are formed, with the same bird returning to the same place night after night. During the day, all the birds disperse across the group's home range. If the land is fairly open, they may stay in small groups. If it is densely wooded, solitary birds are the rule; however, they are never far from another turkey. Contact is maintained by periodic calling. This is the reason hunters use turkey calls; they signal the birds in the area to approach. For most of the year, hens flock with other hens or tend their brood, and gobblers form all-male groups, like the bachelor herds seen in other species. A hen with a young brood usually does not travel too far from her well-hidden nest. She leads the poults out each day to forage, and then returns to the nest at night. Young turkeys are very responsive to their mother's calls. If she utters an alarm, they immediately take cover and become astonishingly hard to find. Mother meanwhile tries to distract the predator, and leads it away from the nest and the young. The poults will remain hidden until she returns and gives a contact call.

The mating system of turkeys is different from that seen in any other domestic bird. Chickens are polygamous; no pair bonds are formed, and mating takes place any time a rooster finds a receptive hen. Turkeys do not form pair bonds either. Their form of courtship is a variation on what is known as a lek system, commonly seen in grouse. During the breeding season, all the adult toms congregate in a central location. Each male selects a small piece of ground, and begins to display. These displays are most impressive, even to a human observer. They involve puffing, whistling, strumming the wings on the ground, and fanning out the tail feathers. Each male slowly marches up and down on his stage, occasionally uttering loud calls. All the females in the area are attracted by these sounds, and approach. Now the males have an audience for their strutting. Eventually, a hen will choose a male, walk up to him and assume a receptive posture. The male stops displaying, walks off his stage, and mating takes place. Usually, a male who has mated that morning will not return to his display ground until the next day, but some will go back right away. Females almost never repeat a mating on the same day. On what basis does a hen choose her male? A lot of people are investigating this topic. Somehow the male manages to communicate his suitability as a mate. There is some evidence from other species that also use this mating system that the location of the stage on which the male displays is a clue to his status. The most dominant males get the central stages, and it appears that females more often select males from these areas. Do they choose these males because the location means the males are "good," or do the males compete for these locations because the females choose central locations? The answer is not yet clear. At any rate, this daily assembly and display continues throughout the breeding season. By the end of this time, the males are physically exhausted and emaciated (they do not take much time to feed), and most of the females are laying eggs. The sexes now separate again, the males to recuperate and the females to incubate (Schorgen, 1966).

If we keep domestic turkeys outdoors where they have sufficient room to move about, we may observe these displays. Of course, some of the strains of domestic turkey have been so selectively bred for broad breasts that they are unable to stand, let alone display and breed. Propagation is then by means of artificial insemination. Also, they tend to go to market before sexual maturity is reached. To watch turkey behavior, select one of the show breeds, such as Bronze or Bourbon Red.

Behavioral problems in the commercial turkey house are similar to those in chicken operations. Turkeys tend to be less aggressive than chickens; the pecking order is not so rigid. However, as we said at the beginning, turkeys are very curious. They will peck at anything. Likewise, they will approach and stare at something that interests them, even if dozens of other turkeys are already there and dozens more are crowding in. Poults have been smothered in large numbers because the whole flock moved over to look at a person standing outside the window. Rules for success are basically the same as for chickens. Be sure there are enough sources of food, water, and heat, and that all are equally accessible. Divide your flocks by age and size. If you are used to working with chickens, do not fret about how much "stupider" turkeys appear. Remember that they are really very different, even if they look similar.

What about domestic ducks? First of all, we must make a distinction between mixed breeds, like the Pekin or the Rouen, and wild-type species that are commonly kept as pets or game birds, such as Mallards or Teal. The Pekin and Rouen are manmade breeds, mixtures of several different breeds created to supply meat. Likewise, the Khaki Campbell is a mixed breed designed to produce quantities of eggs. Mallards, Teal, Wood ducks, Muscovies, and others are naturally occurring species. They are often hand-raised and can become as tame as any manmade breed, but no genetic manipulation has taken place. Of course, if a group of birds becomes too inbred, with no new blood added to the pool, genetic alterations may show up anyway.

The ducks most commonly found on farms and in commercial operations all stem from the Mallard, *Anas platyrhynchos*. These include the domestic Mallard, the large fat Mallard-colored Rouen, and the even larger and fatter white Pekin. The yellow ducklings that appear in pet shops around Easter (often dyed pink and blue) are also Pekins. Mallard and Rouen ducklings look like big bumblebees, downy black and yellow. The other frequently seen farm duck is the Muscovy, *Cairina moschata*. This is a South American tree duck, actually quite dissimilar to the Mallard group in many ways. We will confine our discussion of behavior to *A. platyrhynchos* races, since they are the most likely to be encountered as pets or meat birds.

The wild mallard is a familiar game bird. It is a migratory dabbling duck (as opposed to a diving duck); the males with their striking green heads are popular subjects for wallpaper patterns, neckties, and so on. Not all Mallards migrate. If they live in a temperate zone, where food is available all winter, they will stay there. But if the winters are severe, the flocks take off in the fall, fly to a more southern spot, and return to their breeding and nesting grounds in the spring. (See Chapter 3, Section B on orientation for details on migration.) Mallards exhibit a third type of mating system, pair bonding. Most of the Anatidae (ducks, geese, and swans) show this pattern, but the strength and duration of the bond varies considerably from species to species. Mallards are relatively fickle, as ducks go. Their pair bonds last for only one breeding season, though there is evidence that some pairs seek each other out each year upon arrival at the breeding ground. And these bonds are not observed faithfully by the males!

Three types of mating have been described for Mallards. The first is cooperative pair mating. The female solicits a display from the male with a stereotyped set of signals; the male displays; the female responds with a receptive signal; and mating takes place, usually in the water. After mating, both male and female give "satisfaction" displays. Instead of cooperative pair mating, forced pair mating may occur. In this case, the drake approaches his mate without displaying or waiting for her signal of receptivity, grabs her by the neck feathers and breeds her. The female may protest vigorously and try to escape, or she may appear to give in. At the end, the male usually gives his satisfaction display, and the female may or may not give hers. The third category has been referred to as rape. We prefer not to use this term, as it carries with it a great weight of psychological implications. However, we must admit that its use in this case is not totally inappropriate, at least from the point of view of humans watching the proceedings. In this case, one or more males attempt to breed a female who is not their mate. The female gives every sign of resisting. No displays are made on either side; in fact, this can get very violent, especially if more than one male is involved. Males will chase females on land and water, and even fly after them as they attempt to escape. In some instances, even females on nests are attacked, to the detriment of their eggs. These forced matings by strange males can be avoided if the female's mate is near her. Only lone females are vulnerable.

There are a host of theories about the origins and adaptability of this practice, which is seen in other birds as well. It is to the male's advantage to breed with as many females as possible; this increases his gene pool far more rapidly than if he remains faithful to his mate. On the other hand, whenever he is out chasing another female, some other drake could be doing the same to his mate. The male must weigh his risks and benefits. Will he produce more offspring by staying with his own mate or by chasing others? And what about the female? How can she maximize her own gene pool? In her case, the advantage of multiple matings with different males may be more than offset by the risk of damage to herself and her nest. The pair system is the one best suited to safely raising young. Obviously, Mallard society has evolved to this compromise point. Pair bonds are formed, and kept for the duration of the season, but the males remain opportunistic breeders. No one is too sure about the females. They certainly give every sign of protesting, but some researchers feel that they solicit an occasional out-of-pair mating as well (Ripley, 1957; Johnsgard, 1965). We cannot tell for certain.

This mating system is seen in domestic Mallards and Pekins as well. Male Mallards can be a terror in the breeding season, trying to mate with anything feathered and approximately the right size. Subordinate males are often jumped on if there are not enough females around, as are nesting hens or even geese. Females are often drowned in the pond, as male after male attempts to mate. It is a good idea to try to maintain the sex ratio of your flock at parity if you want to keep your females alive and healthy.

Most people who keep ducks these days do so for other than economic reasons. They make attractive additions to small farms, and most species tame readily. If you can get young ducklings, less than 3 days old, they are easily imprinted onto human beings. (See Fig. 13.) It can be great fun for a child to have a pet duck that follows her about, and it seem to have few lasting effects on the duck's social development. Mallards and Pekins are particularly adaptable to backyard life; we have raised many a brood with a child's wading pool for a pond. Aside from the potential for damage to the females if there are too many males, behavioral management problems in ducks really do not arise.

Domestic geese are less commonly seen than ducks in the United States today. They are also found infrequently in

Figure 13   Many people enjoy the sight of a flotilla of geese on a pond. (Photo by G. Honoré)

the supermarket; there is little demand for goose meat in most of the country, with the exception of a few specialty markets, and larger cities with a European immigrant population. Thus commercial goose farms are almost nonexistent. Contrast this with southern France, where huge flocks of grey Toulouse geese are tended in the fields, and force-fed to ensure large, fat livers. However, geese are seen on U.S. farms, for many of the same reasons as are ducks. Many people raise geese for home consumption; others keep them for their guard dog potential. A recent interview of several convicted tack-room thieves indicated that they preferred to avoid those stables where there were geese. Once the flock is alerted and begins honking, it is impossible to shut it up. A dog may be bribed or drugged with a bit of meat; there is no way to bribe a dozen geese. Aside from practicalities, many people (the authors included) enjoy the sight of a flotilla of geese on the pond, or grazing on the shore.

The most frequently seen domestic geese are the white Emden, the grey Toulouse, and the brown or white Chinese.

Just as with ducks, many breeds are manmade, the result of deliberate or accidental crossbreeding. Toulouse, Emden and other so-called European geese all stem from the wild Greylag (*Anser anser*). The immediate ancestor of the Chinese strains is the Swan goose (*A. cygnoides*), an Asian species. There are numerous other species and subspecies in the genus Anser, many of which are seen migrating in the spring or fall. These wild geese are occasionally kept on waterfowl preserves. Domestic geese tend to be larger and heavier than their wild relatives, characteristics that breeders have selected for. The other wild goose often seen on lakes and fields all over the United States is the Canada goose, *Branta canadensis*. Note that this bird is of a different genus, less closely related to our domestic geese. Though this beautiful bird frequently is available through commercial game farms, it cannot be regarded as a domestic species. Many aspects of its behavior, however, are very similar to those of other geese. (See Fig. 14.)

In fact, it is easy to generalize about goose behavior. Even swans, which we will not discuss, can readily be compared with geese. For the sake of simplicity, we will use the genus

Figure 14 When raised by hand, young precocial birds can make excellent companions. (Photo by P. Klopfer)

Anser as an example, but many of the statements made about it apply equally well to the genus *Branta* and the genus *Cygnus* (swans).

Geese are often used as an example of marital fidelity. They form strong pair bonds lasting as long as the birds live. If one partner dies, the other may or may not take a new mate, depending on its age and a variety of other factors. Naturalists have been fascinated for centuries by the way in which two birds, arriving at the crowded nesting ground, can find one another year after year. Sometimes, of course, they arrive together, but not always. Visual and auditory signals appear to be very important in recogntion. Geese are grazers, and, like other grazing animals, highly social. The flock is a much more significant entity than in ducks. Like sheep, a flock of geese have a leader, one bird who consistently sets the direction of movement. There is no strict dominance hierarchy, except among the young ganders who have not yet chosen a mate, but live in bachelor flocks.

In the wild, goose flocks tend to be family groups, including father, mother, and the young of the past 2 (and sometimes even 3) years. The young females stay with the family longer than do the males. As in many other species of birds and animals, males often travel some distance to find a mate. This reduces the likelihood of too much inbreeding. A young gander may join another flock for a while. Depending on the density of the population, he and his new mate may stay with her flock, or move off on their own. Often two or more families will meet and graze together, always following an elaborate greeting ceremony. It is a peaceful society. Fighting occurs rarely, and then usually among the young ganders in search of mates. Mate selection appears to be a two-way street, in that the appropriate responses of both male and female are essential. Out-of-pair matings are very rare in geese.

When the female is nesting, the gander stands guard. Such a bird can be really fierce, as many curious humans have realized too late. If he is not in the immediate area of the nest, he is within earshot. One cry from his mate, and he will respond at once, rushing back with loud calls. Often he will be accompanied by other young males of the flock. This accessory attendance is seen in domestic geese as well, especially if there are not enough females to go around. Ganders continue their vigilance after the goslings hatch. This is in strong contrast to ducks, where the male's job is done once mating is over. Raising goslings is a true two-parent effort, with older children often still trailing along (Johnsgard, 1968).

All these things may be observed in the barnyard flock. Geese are a little pickier than Pekins; they do need some greenery to feed on. (If no pasture is available, your garden will do nicely.) Their water must also be kept fairly clean to prevent bacterial growth. Aside from these requirements, they are rewarding birds to raise. Their highly social nature leads to a lot of contact between group members. People can readily learn to recognize individual greeting calls and postures, alarm honks, threat hisses and the triumphant trumpet of ganders who have successfully driven off an intruder. The only behavioral problem that geese present is the aggressiveness of the ganders, particularly when the females are on the nest or young are present. Small children can be bruised and frightened by an attacking gander. Even many adults find facing a hissing gander, wings outstretched, to be a daunting experience. However, geese can be taught to respect humans. Many attacks are mostly show; the bird does not actually bite. If one does, it may be necessary to carry a stick for a while, and turn and shout loudly as the gander approaches. Of course, every effort should be made not to disturb the nest or the goslings. Try to think of that nasty gander as a protective parent (and husband) rather than an arrogant nuisance. We feel, after many years of experience, that geese are definitely worth the trouble.

E. Exotic Pets: Cage Birds, Rodents, Reptiles, and Fish

This last category includes many animals that are frequently kept as pets, and therefore show up in veterinary clinics, but they do not fall neatly into any of the groups already discussed. Some of these animals cannot really be considered domesticated, according to the definition we discussed at the beginning of this section. There is no biological justification for lumping all these animals into one category; parakeets bear little resemblance, physically or behaviorally, to pythons. However, they are both wild species, even if captive-bred and quite tame. In many cases, little is known about the natural history or behavior of the animal in question. In others, it has been the subject of extensive research. Thus the format of this section will be slightly different. We will briefly discuss some of the commoner exotic pets, with special attention to aspects of their behavior that are important to their management.

Cage birds have been valued for centuries. Chinese emperors kept nightingales for their singing abilities; great prestige was accorded the man whose bird sang the sweetest. In Indian courts, mynahs were taught to speak for the amusement of the nobles. Everyone knows that parrots were the favorite companions of pirates. Coal miners put their canaries to work; the birds succumbed to lethal gas before the men could detect it, and thus were used as crude telltales. Today, thousands of people enjoy the companionship of a pet bird, from dime-store parakeets to costly macaws.

There are so many species of birds kept as pets that it is impossible to discuss them individually in this chapter. Once again, though, some generalizations may be made. Most cage birds are diurnal; just like their people, they are awake and active during the day and sleep at night. Day for a bird begins as soon as it becomes light, and ends when the light goes. Thus a bird in a room with no windows is totally dependent on artificial lights to signal its day and night. This is important, as most birds will not feed unless it is light. A proper balance of light and dark is also necessary for breeding. Many bird owners have become frustrated by the seeming reluctance of their birds to nest. Often the problem may be traced to too few (or too many) hours of daylight. Many people cover their birdcage at night when the cage is in a room with windows. Thus the bird does not receive the first rays of sun, and stays quiet until its cover is removed. A noisy bird can also be quieted when the cover is put back on. These measures can prove useful in a small apartment with sensitive neighbors.

There are a variety of sources for obtaining cage birds. Reputable dealers will be able to tell you where their birds come from. Some species are still being captured in the wild, and usually imported illegally. It is a very poor idea to buy one of these birds, even aside from the legal question. Many of them carry serious diseases, which affect man as well as other birds. Also, to encourage the capture of wild birds is to encourage the destruction of their population. Several species of parrots are already endangered, and the booming U.S. market for these birds is hastening their extinction by poachers. A good bird dealer will be able to lead you to sources who breed different species in captivity. Thus you can be assured that your bird is healthy and legal, and that you have not contributed to wiping out a species. We strongly urge all potential bird buyers to check carefully on the origins of the birds.

Regardless of where the bird comes from, most new owners want to know how to tame it. This can mean anything from having the canary hop to your finger in the cage to teaching the parrot to talk. The secret to success is the same in any case: consistency and repetition. (See Fig. 15.) In avian flocks, visual and auditory signals are very important, especially repeated stereotypical movements; this is true of pet birds as well. These are species with altricial young, where the young birds remain in the nest and are dependent on the parents for several weeks. Once the young are able to feed themselves, they can be taken from the parents and sent to a new home. Young birds are easier to tame and teach than older ones, simply because they are less set in their ways. Birds learn from repeated experiences, both pleasant and unpleasant. They are truly creatures of habit, and, once used to a routine, are difficult to change. This is the reason that consistency and repetition are so important in training. If you always use the same command and tone of voice, the bird will respond. (We suggest that interested owners go to the library for more detailed references on their particular species.)

Small rodents are also popular pets, especially for children. They are cute and furry, can be handled readily, and fit into compact cages. The attendant odor can be minimized by faithful cleaning and litter removal. The most popular species are mice, *Mus musculus*; hamsters, *Mesocricetus auratus*; gerbils, *Meriones unguiculatus*; and guinea pigs, *Cavia por-*

Figure 15   Patient repetition is the key to training pet birds. (Photo by G. Honoré)

cellus. All of these are domesticated versions of wild species. A variety of color and coat types is available as a result of selective breeding.

Again, generalizations must be made. There are considerable differences between the four species mentioned above, which have been the basis for years of research in physiology, pharmacology, and behavior. Anyone who has kept more than one species as a pet has surely noted this. Yet there are a lot of similarities, and from the point of view of a pet owner, these outweigh the differences. All rodents rely heavily on their olfactory senses for communication. They have well-developed scent organs, and a variety of specialized glands for marking. These are often seen as oily patches on the animal's flanks. They also have a tendency to gnaw on hard surfaces. The big front incisors grow continuously, and must be kept filed down. One of the frequently seen health problems of pet rodents is overgrowth of these teeth. In the wild, constant chewing at roots, tough plants, and fibers keep the teeth aligned. When these animals are kept in plastic or metal cages and given a soft diet, the teeth are not properly worn. Eventually the animal can starve to death, as the overgrown incisors prevent feeding. This is an easy problem to avoid. Recognize the animal's need to gnaw, and provide it with chunks of wood or hard rodent biscuits.

One of the commonest behavioral problems seen in small rodents stems from overcrowding, especially if mixed sexes are present. Some species are more tolerant than others, though eventually all of them reach the point where fighting becomes a serious problem. Rodents do have dominance hierarchies within sexes, and subordinate animals must have room to escape. There are also hierarchies between sexes, but these are more complex. It is essential to keep a close eye on relationships within your colony, especially if you want to raise young. Adult males will often kill newborn pups. So will females, even mothers, if they are stressed. The ideal arrangement is to have no more than one breeding pair in the cage at one time, unless, of course, the cage is particularly spacious and can be subdivided into nooks and crannies. Sometimes it is even necessary to remove the male temporarily when the young are born; there are individual differences that each owner has to discover himself (Hart, 1973). Many books have been written on the subject of rodent care as well; we again suggest a trip to the library.

Some people keep rodents not as pets, but as a food source for large reptiles. The Old and New World constrictors (*Python* and *Boa*) are the most popular of the giant snakes.

Nonvenomous local species, such as the Black Rat Snake
(*Elaphe obsoleta*) and the King Snake (*Lampropeltis getulus*),
are often kept as well. All of these snakes can be frequently
seen in pet stores. Although some serious snake fanciers also
collect and breed certain venomous species, these will not be
discussed here. The risks inherent to this practice limit it to
experienced handlers only. Special techniques and precau-
tions are required. The same is true, of course, of adult
pythons. Many species easily reach 15 feet and 150 pounds in
a few years. That cute little snake you brought home from the
pet shop in a 10-gallon aquarium is now bigger than you are.
For the sake of both the pet and the owner, we strongly urge
first-time snake buyers to consider one of the smaller species.
Certain lizards also adapt well to captive conditions, as long
as their dietary needs can be met. This often means raising
mealworms or other insects, as lizards can be quite finicky.
Special attention must always be paid to the taxonomic identity of the animal. Is it a member of a group that lives in the
moist jungle, or is it a desert species? This is very important
to the animal's health. Even if your individual was bred in
captivity, it has still evolved over countless generations to
adapt to a particular environment. Reptiles are quite sensitive
to changes in temperature and humidity, and these must be
monitored carefully.

The environment has more effect on the behavior of reptiles than on most other animal groups. This is due to their
"cold-bloodedness." In cold weather they become sluggish; in
warm weather, active. Carnival snake handlers have taken
advantage of this effect for years, refrigerating their speci-
mens before the show. Snakes must be fed more often in
warm weather as well. Many temperate species crawl into
dens and hibernate over the winter. If they are kept as pets in
a warm house, however, they will remain active year round.
Veterinarians are frequently approached by reptile owners
whose pets refuse to eat. Once physical illness has been ruled
out, the vet should ask about the conditions in which the
snake (or lizard) is kept. Is it warm enough? Often temperatures that are comfortable for humans are too cool for a snake.
A sluggish reptile will probably not eat, and, if it does, will
have difficulty digesting its meal. Other common problems

are lack of space (that python has outgrown its 10-gallon aquarium) and too much or too little moisture. Any of these factors will affect feeding behavior.

Many snake owners are content to keep their pets in display cages and just look at them. Others want to handle and tame theirs. Can a snake be trained? This is a tricky question to answer, since it depends on what is meant by training. Snakes are certainly capable of discrimination between various signals, both olfactory and visual. Wild snakes maintain territories and home ranges just like other animal groups. In laboratory studies, they respond to noxious stimuli with conditioned avoidance. They also clearly associate other stimuli with the presence of food, and respond accordingly. Obviously, some sort of learning does take place. However, sociality is not highly developed in any reptile group (Shaw and Campbell, 1974; Greenberg and MacLean, 1978). A pet snake will become tame, especially if it is young, with repeated gentle handling. (See Fig. 16.) Snakes bite when frightened or when feeding. If you do not alarm your snake or confuse it with scents of food, it is unlikely to strike. On the other hand, it is asking too much to expect a reptile to respond like a cuddled cat. Tolerance is the most you will get. Snakes lack external ears, so they cannot hear you calling (though they do learn to associate the vibrations of their opening cage door with coming out).

People often wonder if their snake can recognize them. Does the snake know who is holding it? This is really two questions in one. First of all, is the snake capable of discrimination between individual humans? Given the acute olfactory senses of this group, it would be surprising if it were not, though this theory has not been properly tested. Secondly, does the snake associate certain scents with certain individuals? This would be true recognition, which involves not only discrimination but a memory component as well. Again, this has not been tested. It is an interesting question, from an evolutionary as well as a pet owner's point of view.

The most popular pet for apartment dwellers, allergy sufferers and fastidious housekeepers is undoubtably the aquarium fish. Many enthusiasts keep fish by choice, in addition to other pets. They are clean, quiet, relatively easy to care for,

Figure 16   Many snakes will become quite tame and easy to handle. (Photo by G. Honoré)

and a well-kept aquarium can be a beautiful sight. Hundreds of species of fish are available for home aquaria, fresh- and salt-water dwellers, live-bearers and egg-layers, solitary and schooling. Some are tolerant of a wide range of water quality and will eat almost anything, while others must have crystal-clear surroundings and live food. It is up to the prospective buyer to decide how much time (and money) he is willing to put into his aquarium. We urge this buyer to go to the library and consult any one of the several good books on aquarium needs of all fish, to find suggestions about which species are easiest to keep.

Given the enormous range of fish kept in aquaria, it is impossible to make generalizations about their behavior. It is important to know a little about each species you plan to keep, especially if you want to have more than one at a time. Some fish, like the tetras, school readily and feed on small particles. Others, like some of the cichlids, are territorial and carnivorous. Smaller fish will rapidly disappear from the tank if put in with a large Jack Dempsey (*Cichlasoma biocellatum*). This situation is the major example of a behavioral problem in aquarium fish, and is easily avoided.

As with reptiles, the environment in which fish are kept has a profound effect on their behavior. Water temperature and pH are important for more than just the physical health of the fish. So is the topography of the tank. Are you keeping a marine species from a coral reef? Then be sure to include chunks of coral (or at least rocks) in your aquarium design. These fish like to hide in crevices. So do some of the African lake fish. Other lake species prefer a clump of vegetation. Territories are often centered around rocks or large plants. Put in a few of these, and then sit back and watch how the fish use them. It can reveal a lot about their society; who is dominant over whom? Are territories defended against all other fish, or only certain individuals? Are the boundaries fixed, or do they seem to vary? Researchers spend hours in the lab (and in the ocean, and in the lake) trying to answer these questions for different species. Try it! It can add a whole new dimension to your hobby.

Epilogue

By this point the reader should have a good grasp of the principles of animal behavior and be able to apply these to different animal groups. (See Fig. 17) We hope that some common themes have come through. An understanding, or at least a glimpse of an understanding, of why an animal acts as it does increases our appreciation of that animal. We can more easily care for it to its best advantage, and anticipate its needs. Whether we are veterinarians, farmers, pet owners, or observers, our lives are to varying degrees tied to the lives of animals. Often we are in control of those lives. Comprehension and compassion tend to go together, and both are an essential part of any education.

Figure 17   Curiosity and mutual respect between members of two species; a feral foal approaches the photographer's dog. (Photo by B. Franke Stevens)

Bibliography

Ardrey, R. (1966). "The Territorial Imperative." Atheneum, New York.

Axelrod, R. (1985). "The Evolution of Cooperation." Basic Books, New York.

Barlow, J. (1964). Inertial navigation as a basis for animal navigation. *J. Theor. Biol.* **6,** 76–117.

Bartholomew, G., and Birdsell, J. (1953). Ecology and the Protohominids. *Am. Anthropol.* **55,** 481–498.

Brown, J. L. (1964). The evolution of diversity in avian territorial systems. *Wilson Bull.* **76,** 160–169.

Brown, J. L. (1975). "The Evolution of Behavior." Norton, New York.

Chevalier-Skolnikoff, S., and Poirier, F. E., eds. (1977). "Primate Bio-social Development." Garland, New York.

Cloudsley-Thompson, I. L. (1961). "Rhythmic Activity in Animal Physiology and Behavior." Academic Press, New York.

Commoner, B. (1971). "The Closing Circle." Knopf, New York.

Craig, J. V. (1981). "Domestic Animal Behavior: Causes and Implications for Animal Care and Management." Prentice-Hall, Englewood Cliffs, New Jersey.

Czeisler, C. A., *et al.* (1989). Bright light induction of strong resetting of the human circadian pacemaker. *Science* **244,** 1328–1332.

Darling, F. F. (1937). "A Herd of Red Deer." Cambridge Univ. Press, London.

Darwin, C. (1876). "The Movements and Habits of Climbing Plants." Appleton, New York.

Deegener, P. (1918). "Die Formen der Vergesellschaftung im Tierreiche." Veit, Leipzig, German Democratic Republic.

Drews, D. (1973). Group formation in captive g.c.: Notes on the dominance concept. Z. Tierpsychol. **32,** 425–435.

Eaton, R. L. (1974). "The Cheetah." Krieger, Malabar, Florida.

Ellis, D. V. (1985). "Animal Behavior and Its Applications." Lewis, Chelsea, Michigan.

Emlen, S. (1969). Bird migration: Influence of physiological state upon celestial orientation. Science **165,** 716–718.

Espinas, A. (1878). "Des Societes Animales." Baillière, Paris.

Estes, R. D. (1974). Social organization of the African Bovidae. In "Behavior of Ungulates and Its Relation to Management" (V. Geist and F. Walther, eds.) Gresham, Surrey, England.

Exner, S. (1893). Negativ Versuchsergebnisse ueber das Orientierungsvermoegen der Brieftauben. Sitzungsber. Akad. Wiss. Wien, Math.–Naturwiss. Kl., Abt. 3 **102,** 318–331.

Fagen, R. (1981). "Animal Play Behavior." Oxford Univ. Press, Oxford, England.

Folley, S. J., and Kraggs, G. S. (1965). Oxytocin levels in the blood of ruminants with special reference to the milking stimulus. In "Symposium on Advances in Oxytocin Research, 1964" (J. Pinkerton, ed.), 37–49. Pergamon, Oxford, England.

Fox, M. W. (1974). "Understanding Your Cat." Coward-McCann, New York.

Fox, M. W., ed. (1975). "The Wild Canids: Their Systematics, Behavioral Ecology and Evolution." Van Nostrand-Reinhold, New York.

Fuller, J. L. (1969). The genetics of behaviour. In "The Behaviour of Domestic Animals" (E. S. S. Hafez, ed.), 2nd Ed. Williams & Wilkins, Baltimore, Maryland.

Gallup, G. G., Jr. (1970). Chimpanzees: Self-recognition. Science **167,** 86–87.

Geist, V., (1971). "Mountain Sheep: A Study in Behavior and Evolution." Univ. of Chicago Press, Chicago, Illinois.

Golani, I. (1973). Non-metric analysis of behaviour interaction sequences in captive jackals. *Behaviour* **44,** 89–112.

Gottlieb, G. (1970). "Development of Species Identification in Birds." Univ. of Chicago Press, Chicago, Illinois.

Grandin, T. (1978). Design of lairage, yard and race systems for handling cattle in abattoirs, auctions, ranches, restraining chutes and dipping vats. *Proc. World Congr. Ethol. Appl. Zootechnol. 1st.*

Greenberg, N., and MacLean, P. D. eds. (1978). "Behavior and Neurology of Lizards: An Interdisciplinary Colloquium." Natl. Inst. Mental Health, Bethesda, Maryland.

Griffin, D. R. (1944). The sensory basis of bird migration. *Q. Rev. Biol.* **19,** 21–32.

Griffin, D. R. (1952). Bird navigation. *Biol. Rev.* **27,** 359–400.

Griffin, D. R. (1958). "Listening in the Dark." Yale Univ. Press, New Haven, Connecticut.

Griffin, D. R. (1976). "The Question of Animal Awareness: Evolutionary Continuity of Mental Experience." Rockefeller Univ. Press, New York.

Hafez, E. S. W., ed. (1975). "The Behaviour of Domestic Animals," 3rd Ed. Williams & Wilkins, Baltimore, Maryland.

Hailman, J. P. (1967). "Ontogeny of an Instinct. Behaviour Supplement." Brill, Leiden, The Netherlands.

Hale, E. B. (1969). Domestication and the evolution of behaviour. *In* "The Behaviour of Domestic Animals" (E. S. E. Hafez, ed.), 2nd Ed. Williams & Wilkins, Baltimore, Maryland.

Hall, W. G., and Williams, C. L. (1983). Suckling isn't feeding, or is it? *Adv. Study Behav.* **13.**

Hamilton, W. D. (1963). The evolution of altruistic behavior. *Am. Nat.* **97,** 354–356.

Harkness, J. E., and Wagner, J. E. (1983). "The Biology and Medicine of Rabbits and Rodents," 2nd Ed. Lea & Febinger, Philadelphia, Pennsylvania.

Hart, M. (1973). "Rats." Allison & Busby, London.

Hess, E. H. (1956). Space perception in the chick. *Sci. Am.* **195,** 71–80.

Hinde, R. (1966). "Animal Behaviour." McGraw-Hill, New York.

Horn, G., Rose, S., and Bateson, P. (1973). Experience and plasticity in the nervous system. *Science* **181**, 506–514.

Howard, E. (1920). "Territory in Bird Life." Murray, London.

Huntingford, F. (1984). "The Study of Animal Behaviour." Chapman & Hall, London.

Jaynes, J. (1969). The historical origins of "ethology" and "comparative psychology." *Anim. Behav.* **17**, 601–606.

Johnsgard, P. (1965). "Handbook of Waterfowl Behaviour." Cornell Univ. Press, Ithaca, New York.

Johnsgard, P. (1968). "Waterfowl." Univ. of Nebraska Press, Lincoln, Nebraska.

Joslin, J., Feltcher, H., and Emlen, J. (1964). A comparison of the responses to snakes of lab- and wild-reared rhesus monkeys. *Anim. Behav.* **12**, 348–353.

Kagan, J., and Beach, F. H. (1953). Effects of early experience on mating behavior in male rats. *J. Comp. Physiol. Psychol.* **46**, 204–208.

Keeton, W. T. (1971). Magnets interfere with pigeon homing. *Proc. Natl. Acad. Sci. U.S.A.* **68**, 102–106.

Kiepenheuer, J. (1984). The magnetic compass mechanism of birds and its possible association with the shifting course directions of migrants. *Behav. Ecol. Sociobiol.* **14**, 81–99.

Kiley, M. (1974). Behavioural problems of some captive and domestic ungulates. *In* "Behaviour of Ungulates and Its Relation to Management" (V. Geist and F. Walther, eds.), Gresham, Surrey, England.

King, A. P., and West, M. (1988). Searching for the functional origins of song in eastern brown-headed cowbirds, *Molothrus ater ater. Anim. Behav.* **36**, 1575–1588.

Kipling, R. (1902). "Just So Stories." Durst, New York (1978 reproduction).

Klama, J. (1988). "Aggression: The Myth of the Beast Within." Wiley, New York.

Klingel, H. (1974). A comparison of the social behavior of the Equidae. *In* "Behaviour of Ungulates and Its Relation to Management" (V. Geist and F. Walther, eds.), Gresham, Surrey, England.

Klopfer, P. H. (1969). "Habitats and Territories." Basic Books, New York.

Klopfer, P. H. (1971). Mother love: What turns it on? *Am. Sci.* **59,** 404–407.

Klopfer, P. H. (1974). "An Introduction to Animal Behavior: Ethology's First Century." Prentice-Hall, Englewood Cliffs, New Jersey.

Klopfer, P. H., and Polemics, J. (1989). Have animals rights? *J. Elisha Mitchell Soc.* **104,** 99–107.

Kramer, G. (1952). Experiments on bird orientation. *Ibis* **94,** 265–285.

Kropotkin, P. (1914). "Mutual Aid, a Factor in Evolution." Knopf, New York.

Lack, D. (1954). "The Natural Regulation of Animal Numbers." Oxford Univ. Press, Oxford, England.

Lehrman, D. S. (1965). Interaction between internal and external environments in the regulation of the reproductive cycle of the ring dove. *In* "Sex and Behavior" (F. A. Beach, ed.), pp. 355–380. Wiley, New York.

Lessley, B. (1990). Differing visions of the animal–human relationship. *Perspect. Ethol.* **9.**

Lewontin, R., Rose, S., and Kamin, L. (1984). "Not in Our Genes." Penguin, New York.

Leyhausen, P. (1979). "Cat Behavior." Garland, New York.

Loeb, J. (1918). "Forced Movements, Tropisms, and Animal Conduct." Lippincott, Philadelphia, Pennsylvania.

Lorenz, K. Z. (1952). "King Solomon's Ring." Crowell, New York.

Lorenz, K. Z. (1974). Analogy as a source of knowledge. *Science* **185,** 229–234.

Marler, P. (1963). Inheritance and learning in the development of animal vocalizations. *In* "Acoustic Behavior of animals" (R. Busnel, ed.). Elsevier, New York.

Mason, J. W. (1968). A review of psychoendocrine research on the pituitary adrenal cortical system. *Psychosom. Med.* **30,** 576–607.

Matthews, G. V. T. (1955). "Bird Navigation." Cambridge Univ. Press, New York.

Medawar, P. (1957). "The Uniqueness of the Individual." Methuen, London.

Mittlestaedt, H. (1962). Control systems of orientation in insects. *Annu. Rev. Entomol.* **7,** 127–198.

Mloszewski, M. J. (1983). "The Behavior and Ecology of the African Buffalo." Cambridge Univ. Press, New York.

Monson, G., and Sumner, L., eds. (1980). "The Desert Bighorn." Univ. of Arizona Press, Tucson, Arizona.

Neilhart, J. G. (1959). "Black Elk Speaks." Pocket Books, New York.

Nievergelt, B. (1974). A comparison of rutting behavior and grouping in the Ethiopian and Alpine ibex. *In* "Behaviour of Ungulates and its Relation to Management" (V. Geist and F. Walther, eds.). Gresham, Surrey, England.

Nottebohm, F. (1970). Ontogeny of bird song. *Science* **167**, 950–956.

Oyama, S. (1985). "The Ontogeny of Information." Cambridge Univ. Press, New York.

Paton, W. (1984). "Man and Mouse: Animals in Medical Research." Oxford Univ. Press, Oxford, England.

Pedersen, C. A., and Prange, H. G. (1979). Induction of maternal behavior in virgin rats after intracerebroventricular administration of oxytocin. *Proc. Natl. Acad. Sci. U.S.A.* **12**, 6661–6665.

Peeters, G., Debackere, M., Lauryssens, M., and Kuhn, E. (1965). Studies in the release of oxytocin. *In* "Symposium on Advances in Oxytocin Research, 1964" (J. Pinkerton, ed.). Pergamon, Oxford, England.

Pond, W. G., and Houpt, K. H. (1978). "The Biology of the Pig." Cornell Univ. Press, Ithaca, New York.

Pribram, K. H., and Melges, F. T. (1969). Psychophysiological basis of emotion. *In* "Handbook of Clinical Neurology" (P. J. Vinken and G. W. Brwyn, eds.). Wiley (Interscience), New York.

Protsch, R., and Berger, R. (1973). Earliest radiocarbon dates for domesticated animals. *Science* **179**, 235–239.

Ripley, F. D. (1957). "A Paddling of Ducks." Harcourt Brace, New York.

Schjelderup-Ebbe, T. (1922). Beitrage zur Socialpsychologie des Haushuhns. *Z. Tierpsychol.* **88**, 225–252.

Schmidt-Koenig, K. (1975). "Migration and Homing in Animals." Springer, New York.

Schmidt-Koenig, K. (1979). "Avian Orientation and Navigation." Cambridge Univ. Press, New York.

Schmidt-Koenig, K., and Ganzhern, J. (1990). On the problem of bird navigation. *Perspect. Ethol.* **9.**

Schnierla, T. C. (1953). Basic problems in the nature of insect behavior. *In* "Insect Physiology" (K. D. Roedes, ed.), pp. 656–684. Wiley, New York.

Schorger, A. W. (1966). "The Wild Turkey." Univ. of Oklahoma Press, Norman, Oklahoma.

Scott, J. P., and Fuller, J. L. (1965). "Genetics and Social Behavior of the Dog." Univ. of Chicago Press, Chicago, Illinois.

Shaw, C. E., and Campbell, S. (1974). "Snakes of the American West." Knopf, New York.

Stresemann, E. (1951). "Die Entwicklung der Ornithologie." Peters, Berlin.

Thorpe, W. H. (1961). "Bird-Song." Cambridge Univ. Press, London.

Thorpe, W. H. (1979). "Origin and Rise of Ethology." Praeger, New York.

Tinbergen, N. (1951). "The Study of Instinct." Oxford Univ. Press, New York.

Tinbergen, N. (1958). "Curious Naturalists." Country Life, London.

Trivers, R. (1985). "Social Evolution." Cummings, Menlo Park, California.

Tyler, S. (1972). The behaviour and social organization of the new forest ponies. *Anim. Behav. Monogr.* **5,** 85–196.

von Frisch, K. (1955). "The Dancing Bees." Harcourt Brace Jovanovich, New York.

von Uexküll J. (1909). "Umwelt und Innenwelt der Tiere." Springer-Verlag, Berlin.

Waring, G. (1983). "Horse Behavior." Noyes, Park Ridge, New Jersey.

Watson, J. B., and Lashley, K. S. (1915). An historical and experimental study of homing. *Carnegie Inst. Washington Publ.* **211,** 7–60.

Wecker, S. C. (1963). The role of early experience in habitat selection by the prairie deermouse, *Peromyscus maniculatus bairdi. Ecol. Monogr.* **33,** 307–325.

Wells, M. J. (1959). Functional evidence for neuronal fields representing the individual arms within the central nervous system of Octopus. *J. Exp. Biol.* **36,** 501–511.

Wells, M. J. (1962). "Brain and Behavior in Cephalopods." Stanford Univ. Press, Stanford, California.

Wenner, A. M. (1967). Honeybees: Do they use the distance information contained in their dance maneuver? *Science* **155,** 847–849.

Wilson, E. O. (1975). "Sociobiology." Belknap, Cambridge, Massachusetts.

Wood-Gush, D. G. M. (1971). "The Behaviour of the Domestic Fowl." Heinemann, London.

Wood-Gush, D. G. M. (1983). "Elements of Ethology." Chapman & Hall, London.

Wrogemann, N. (1975). "Cheetah under the Sun." McGraw-Hill, New York.

Yeagley, H. L. (1947). A preliminary study of a physical basis of bird navigation. *J. Appl. Phys.* **18,** 1035–1063.

Zeuner, F. E. (1963). "A History of Domesticated animals." Hutchinson, London.

Supplemental Reading List

It is often useful to have available a reading list of relatively nontechnical, yet authoritative, articles with which to gain a firmer hold on a subject. Therefore, in addition to the references provided at the end of chapters 1, 2, and 3, there follows a selected list of recent (1950–1984) articles published in *Scientific American*, arranged according to the causes and origins of behavior.

I. General Behavioral Description and Behavioral Maintenance

Milne, L. J. and Milne, M. J. Animal courtship, July 1950.
Tinbergen, N. Curious behavior of the stickleback, Dec. 1952.
Steinmen, D. B. Courtship of animals, Nov. 1954.
Guhl, A. M. Social order of chickens., Feb. 1956.
Marshall, A. J. Bowerbirds, June 1956.
Tinbergen, N. Defense by color, Oct. 1957.
Kind, J. A. Social behavior of prairie dogs, Oct. 1959.
Washburn, S. L. and DeVore, I. Social life of baboons, June 1961.
Eibl-Eibesfeldt, I. Fighting behavior of animals, Dec. 1961.
Calhoun, J. B. Population density and social pathology, Feb. 1962.
von Frisch, K. Dialects of the bees, Aug. 1962.

Wenner, A. M. Sound communication in honeybees, April 1964.

Emlen, J. E. and Penney, R. L. The navigation of penguins, Oct. 1966.

Singh, D. Urban monkeys, July 1969.

Bennet-Clark, H. C. and Ewing, A. W. The love song of a fruit fly, July 1970.

Todd, J. M. The chemical language of fishes, May 1971.

Wilson, E. O. Animal communication, Sept. 1972.

Premack, A. J. and Premack, D. Teaching language to an ape, Oct. 1972.

Topoff, H. R. The social behavior of army ants, Nov. 1972.

Teleki, G. The omnivorous chimpanzee, Jan. 1973.

Thorpe, W. H. Duet-singing birds, Aug. 1973.

Rothschild, M., *et al.*, The flying leap of the flea, Nov. 1973.

Bertram, B. C. R. The social systems of lions, May 1975.

Wilson, E. O. Slavery in ants, June 1975.

Burgess, J. W. Social spiders, Mar. 1976.

Milne, L. J. and Milne, M. The social behavior of burying beetles, Aug. 1976.

Eaton, G. G. The social order to Japanese Macaques, Oct. 1976.

Bekoff, M. and Wells, M. C. The social ecology of coyotes, April 1980.

Lloyd, J. E. Mimicry in the sexual signals of fireflies, July 1981.

Partridge, B. L. The structure and function of fish schools, June 1982.

Ligon, J. D. and Ligon, S. H. The cooperative breeding behavior of the Green Woodhoopoe, July 1982.

Seeley, T. D. How honeybees find a home, Oct. 1982.

II. Behavioral History

Munn, N. L. The evolution of mind, June 1957.

Lorenz, K. Z. The evolution of behavior, Dec. 1958.

Tinbergen, N. Evolution of behavior in gulls, Dec. 1960.

Gilliard, E. T. The evolution of bowerbirds, Aug. 1963.

Wynne-Edwards, V. C. Population control in animals, Aug. 1964.

Bitterman, M. E. The evolution of intelligence, Jan. 1965.

Andrew, R. J. The origins of facial expressions, Oct. 1965.

Esch, H. The evolution of bee language, April 1967.

Seilacher, A. Fossil behavior, Aug. 1967.

Benzer, S. Genetic dissection of behavior ("mosaic" fruit fly genetics), Dec. 1973.

Maynard Smith, J. The evolution of behavior, Sept. 1978.

Bergerund, A. T. Prey switching in a simple ecosystem, Dec. 1983.

Scheller, R. H. and Axel, R. How genes control an intimate behavior, March 1984.

Horner, J. R. The nesting behavior of dinosaurs, April 1984.

III. Behavioral Ontogeny

Warden, C. J. Animal Intelligence, June 1951.

Skinner, B. F. How to teach animals, Dec. 1951.

Butler, R. A. Curiosity in monkeys, Feb. 1954.

Pastore, N. Learning in the canary, June 1955.

Hess, E. H. Imprinting in animals, March 1958.

Harlow, H. F. Love in infant monkeys, June 1959.

Gibson, E. J. and Walk, R. D. The "visual cliff", April 1960.

Dilger, W. C. Behavior of lovebirds, Jan. 1962.

Harlow, H. F. and Harlow, M. K. Social deprivation in monkeys, Nov. 1962.

Best, J. B. Protopsychology, Feb. 1963.

Denenberg, V. H. Experience and emotional development, June 1963.

Wecker, S. C. Habitat selection, Oct. 1964.

Hailman, J. P. How an instinct is learned, Dec. 1969.

Rosenzweig, M. R., Bennett, E. L., and Diamond, M. L. Brain changes in response to experience, Feb. 1972.

Geschwind, N. Language and the brain, April 1972.

Hess, E. H. "Imprinting" in a natural laboratory, Aug. 1972.

Rosenblatt, J. S. Learning in newborn kittens, Dec. 1972.

Salk, L. The role of the heartbeat in the relations between mother and infant, May 1973.

R. Menzel, and Erber, J. Learning and memory in bees, July 1978.

Alkon, D. L. Learning in a Marine Snail, July 1983.

IV. Behavioral Control

Griffin, D. R. Navigation of bats, Aug. 1950.
Kalmus, H. Sun navigation of animals, Oct. 1954.
Sperry, R. W. Eye and the brain, May 1956.
Hess, E. H. Space perception of the chick, July 1956.
Guttman, N. and Kalish, H. I. Experiments in discrimination, Jan. 1958.
Griffin, D. R. More about bat "radar," July 1958.
Grundfest, H. Electric fishes, Oct. 1960.
Blough, D. S. Animal psychophysics, July 1961.
von Holst, E. and St. Paul, U. Electrically controlled behavior, March 1962.
Lissman, H. N. Electric location by fishes, March 1963.
Wilson, E. O. Pheromones, May 1963.
Hubel, D. H. The visual cortex of the brain, Nov. 1963.
Muntz, W. R. A. Vision in frogs, March 1964.
Lehrman, D. S. The reproductive behavior of ring doves, Nov. 1964.
MacNichol, Jr., E. F. Three-pigment color vision, Dec. 1964.
Boycott, B. B. Learning in the octopus, March 1965.
Roeder, K. D. Moths and ultrasound, April 1965.
Held, R. Plasticity in sensory-motor systems, Nov. 1965.
Peterson, L. R. Short-term memory, July 1966.
Agronoff, B. W. Memory and protein synthesis, June 1967.
Wilson, D. M. The flight control system of the locust, May 1968.
Pribram, K. The neurophysiology of remembering, Jan. 1969.
Greenwalt, C. How birds sing, Nov. 1969.
Lucia, A. R. The functional organization of the brain, March 1970.
Kandel, E. R. Nerve cells and behavior, July 1970.
Levine, S. Stress and behavior, Jan. 1971.
Atkinson, R. C. and Shiffrin, R. M. The control of short-term memory, Aug. 1971.
Harris, J. F. The infrared receptors of snakes, May 1973.
Schneider, D. The sex-attractant receptor of moths, July 1974.

Bentley, D. and Hoy, R. The neurobiology of cricket song, Aug. 1974.

Keeton, W. T. The mystery of pigeon homing, Dec. 1974.

Emlen, T. The Stellar-orientation system of a migratory bird, Aug. 1975.

Saunders, D. S. The biological clocks of insects, Feb. 1976.

Crews, D. The hormonal control of behavior in a lizard, Aug. 1979.

Knudsen, E. I. The hearing of the barn owl, Dec. 1981.

Shettleworth, S. J. Memory in food-hoarding birds, Mar. 1983.

Bronson, F. H. The adaptability of the house mouse, Mar. 1984.

Moore, J. Parasites that change the behavior of their hosts, May 1984.

Index